Affective Aspects of Learning

European University Studies
Europäische Hochschulschriften
Publications Universitaires Européennes

Series VI
Psychology

Reihe VI Série VI
Psychologie
Psychologie

Vol./Bd. 759

PETER LANG
Frankfurt am Main · Berlin · Bern · Bruxelles · New York · Oxford · Wien

Junmei Xiong

Affective Aspects of Learning

Adolescents' Self-Concept, Achievement Values, Emotions, and Motivation in Learning Mathematics

PETER LANG
Internationaler Verlag der Wissenschaften

Bibliographic Information published by the Deutsche Nationalbibliothek
The Deutsche Nationalbibliothek lists this publication in the Deutsche Nationalbibliografie; detailed bibliographic data is available in the internet at <http://www.d-nb.de>.

Zugl.: München, Univ., Diss., 2008

D 19
ISSN 0531-7347
ISBN 978-3-631-58668-6

© Peter Lang GmbH
Internationaler Verlag der Wissenschaften
Frankfurt am Main 2009
All rights reserved.

All parts of this publication are protected by copyright. Any utilisation outside the strict limits of the copyright law, without the permission of the publisher, is forbidden and liable to prosecution. This applies in particular to reproductions, translations, microfilming, and storage and processing in electronic retrieval systems.

www.peterlang.de

Acknowledgments

I am sincerely thankful for the insightful guidance of my supervisor, Prof. Dr. Rudolf Tippelt, the continuous support from my family, and the patient assistance from my colleagues and friends. Without their guidance, support, and assistance, neither meaningful results would have been found nor could the manuscript have been presented here.

I sincerely thank Prof. Dr. Ditton and Prof. Dr. Fischer for evaluating the dissertation, Prof. Dr. Tock Keng Lim for her comment and suggestions on the statistical part of the dissertation and Dr. Hans Steinmueller for his assistance with the abstract translation from English to German. The names of the colleagues and friends who assisted me with the data collection should be mentioned here, too. They are Dr. Xue Yongling of Peking University, Mr. Zhao Daheng of Peking No Eight Secondary School, M.A. Mr. Qin Yi of China University of Geosciences, M.A. Ms. Jiao Xueli of Munich University, Ms. Zhao Jun, Ms. Gu Xiaoqing, Ms. Cui Reizhen, Mr. Lian Li, Mr. Zhang Qingfeng of Shi Jiazhuang No Two Senior High School, Ms. Zhan Yanling of Wuhan No Sixteen Senior High School, Ms. Chen Li of Shui Guohu Senior High School. I also want to thank the participants of the study, the more than four hundred teenagers from three senior high schools across two cities in China. The majority of them took part in the study with interest and described their academic and emotional experiences in learning mathematics and perceptions of teaching and learning in the school context as accurately as they could during the survey.

I am grateful for my father, Xiong Chengwang, who encouraged me to explore the world without fear, and my mother, Li Jizhen, who sent her only child me to thousand miles away to pursue what I believed important in my life. It was an especially hard decision for her since my father passed away when I was a teenager. I am also grateful for the support and understanding of my husband, Qin Zikai, and the jingle and laughter of my four year old son, Qin Jingpeng, who brings the sunshine to me when it looks dark and cloudy.

Through designing, field testing, analysizing, and integrating processes, I realized that the product of individual human endeavor is always based upon collective effort and therefore it is the researcher's responsibility to be true to science and investigate real life problems for the benefit of the people who are troubled by the problems and wait for suggestions and clarification. It was an initial step for me to try to understand adolescents' emotion development and its functions. There are much more to be done in the future. The little page covers my sincere dedication to understanding and helping youth.

Junmei Xiong
Munich, 9th of January, 2009

Zusammenfassung[1]

Diese Studie beschreibt und analysiert die affektiven und kognitiven Prozesse des Mathematik-Lernens bei Jugendlichen. Die empirische Grundlage der Studie ist eine Umfrage die mit 464 Schülern der 11. Klasse in zwei Großstädten in der Volksrepublik China durchgeführt wurde. Die Fragebogen wurden anhand des "Motivated Strategies for Learning Questionnaire" von Pintrich et al (1991) und des "Academic Emotions Questionnaire – Mathematics" von Pekrun et al (2005) entworfen.

Statistische Korrelation und Regression zeigt dass positive Emotionen (Freude und Stolz) positiv mit der Anwendung von Lernstrategien (wie Üben, Evaluierung, Organisation, kritisches Denken, meta-kognitive Selbst-Regulierung) korrelieren; dahingegen korrelieren negative Emotionen (Zorn, Angst, Scham, Hoffnungslosigkeit, und Langeweile) negative mit der Anwendung von Lernstrategien im Allgemeinen. Selbsteinschätzung von mathematischen Fähigkeiten und die Bewertung von Leistung korrelieren positiv mit positiven Emotionen und der Anwendung von Lernstrategien; sie korrelieren hingegen negative mit negativen Emotionen, mit der Ausnahme das extrinsischer Wert positiv mit Angst und Scham korreliert.

Weiterhin wurde festgestellt dass die Selbsteinschätzung von mathematischen Fähigkeiten signifikanten Einfluss auf die Beziehung zwischen positiven Umweltfaktoren (Aufmerksamkeit während des Unterrichts, das Verhalten gegenüber Mitschülern, der Enthusiasmus der Lehrer) und positiven Emotionen hat. Ebenso hat die Selbsteinschätzung von mathematischen Fähigkeiten Einfluss auf die Beziehung zu negative Umweltfakoren (Kontrolle durch die Eltern, strenge Evaluierung, normativer Fokus in der Evaluierung) und negativen Emotionen. Es wird auch gezeigt dass Aufsuchenziele (approach goals) eine vermittelnde Funktionen haben was den Einfluß von positiven Emotionen auf die Anwendung von Lernstrategien betrifft; Meidenziele (avoidance goals) haben dagegen keine Funktion in Beziehung mit dem Einfluß von Emotionen auf Lernstrategien.

Multiple Regressionsanalyse zeigt dass Langeweile (negatives Prädiktor), Kontrolle durch die Eltern (negatives Prädiktor), meta-kognitive Selbst-Regulierung, Selbsteinschätzung von mathematischen Fähigkeiten, performative Aufsuchenziele (Performance approach goals), Übung (negatives Prädiktor), und strenge Evaluierung (negatives Prädiktor) Prädiktoren des Erfolgs im Lernen von Mathematik sind.

Dieser Erfolg ist besonders gefährdet durch negative Umweltfaktoren und negative Gefühlszustände. Weil Selbsteinschätzung von mathematischen Fähigkeiten die Auswirkungen von negativen Umweltfaktoren beim Lernen auf negative Emotionen abschwächen kann, ist es sehr wichtig das Selbsteinschätzung von mathematischen Fähigkeiten von Schülern aufrecht zu erhalten und zu fördern, sowohl in der Schule als auch zuhause. Deshalb wird weniger Kontrolle durch die Eltern und mehr Eigenständigkeit, weniger strenge und weniger normative Evaluierung, und die Förderung der Selbsteinschätzung von mathematischen Fähigkeiten von Schülern wird empfohlen.

Der Erfolg beim Mathematik-Lernen ist eine Funktion der affektiven und kognitiven Prozesse und deshalb ein multidimensionales Konstrukt. Experimentelle Studien, beispielsweise die statistische Manipulation der Gefühle von Schülern beim Lernen und die Beobachtung der Veränderungen bei der Anwendung von verschiedenen Lernstrategien, werden vorgeschlagen um zu weiteren Schlussfolgerungen über die affektiven und kognitiven Prozesse im Mathematik-Lernen bei Jugendlichen zu gelangen.

Schlagwörter: Kognition und Emotion, Lernstrategien, Mathematik, Selbsteinschätzung von mathematischen Fähigkeiten (math self-concept), Bewertung von Leistung, Leistungsziele, Umweltfaktoren

Abstract[1]

This empirical study describes and analyzes adolescents' affective and cognitive processes involved in learning mathematics. The empirical basis is a survey done with students of grade 11 in two major cities in the People's Republic of China; the survey's questionnaire is based on the "Motivated Strategies for Learning Questionnaire" by Pintrich et al (1991) and the "Academic Emotions Questionnaire – Mathematics" by Pekrun et al (2005).

Correlation-regression statistics revealed that positive emotions (enjoyment and pride) positively correlated with learning strategy use (rehearsal, evaluation, organization, critical thinking, and metacognitive self-regulation), whilst negative emotions (anger, anxiety, shame, hopelessness, and boredom) negatively correlated with learning strategy use in general; math self-concept and achievement values positively correlated with positive emotions and learning strategy use, and negatively correlated with negative emotions except that extrinsic value positively correlated with anxiety and shame.

Math self-concept was found to mediate the impact of positive environmental factors (lecture engagement, peer attitude, teacher enthusiasm) on positive emotions; math self-concept was also found to mediate the impact of negative environmental factors (parental control, harsh evaluation, normative evaluation focus) on negative emotions. Furthermore, the approach goal was found to mediate positive emotions' impact on learning strategy use; the avoidance goal did not function as a mediator between emotions and learning strategies.

According to multiple regression analysis, boredom (negative predictor), parental control (negative predictor), metacognitive self-regulation, math self-concept, performance approach goal, rehearsal (negative predictor), and harsh evaluation (negative predictor) were predictors of math achievement.

Math achievement is compromised in particular by negative environmental factors and negative affect. Since math self-concept could buffer negative learning environment's impact on negative affect, maintaining and enhancing students' math self-concept is important in both school and home contexts. Therefore, less parental control and more autonomy, less harsh evaluation and normative evaluation are recommended.

Achievement in mathematics is a function of affective-cognitive processes and thus a multidimensional construct. Experimental studies, for example manipulating students' affect in learning to a certain degree and observing their changes in learning strategy use, are suggested to reach further conclusions regarding the affective and cognitive processes in learning mathematics.

Key words: cognition and emotion, learning strategies, mathematics, math self-concept, achievement values, achievement goals, environmental factors, etc.

1 Paper presented at the XXIX International Congress of Psychology, 21 - 25 July, 2008, Berlin, Germany.

Our most important are our earliest years.

William Cowper (1731-1800), English author

Table of Contents

LIST OF TABLES		17
LIST OF FIGURES		20
LIST OF DIAGRAMS		21

Chapter I   Introduction ---------- 23
1.1   **Purpose of the study** ---------- 23
    1.1.1   Adolescence – a critical point ---------- 24
    1.1.2   Learning mathematics – cognitive functioning with affect ---------- 27
1.2   **Collectivist and individualist cultures and implications for motivation** ---------- 28
1.3   **Cognitive and motivational characteristics of different groups in math learning context** ---------- 29
    1.3.1   High versus low achievement groups ---------- 29
    1.3.2   Male versus female groups ---------- 29
1.4   **Research questions** ---------- 32
1.5   **School system in China** ---------- 33

Chapter II   Theoretical Background ---------- 35
2.1   **Self-esteem and self-concept** ---------- 35
    2.1.1   Self-esteem ---------- 35
    2.1.2   Self-concept ---------- 38
        2.1.2.1   Origins and development of self-concept ---------- 38
        2.1.2.2   Subjective and objective selves ---------- 39
        2.1.2.3   Neisser's model of tripartite self ---------- 40
        2.1.2.4   Self-concept and identity diffusion ---------- 41
        2.1.2.5   Impact on achievement ---------- 42
    2.1.3   Hypotheses on math self-concept ---------- 43
2.2   **Achievement motivation** ---------- 43
    2.2.1   Definitions ---------- 43
    2.2.2   Development of achievement motivation ---------- 44
        2.2.2.1   Three-stage development ---------- 44
        2.2.2.2   Childrearing practices and strength of motives ---------- 44
    2.2.3   Development of achievement motivation theories ---------- 45
        2.2.3.1   Achievement motivation according to Murray ---------- 45
        2.2.3.2   Expectancy-value theory of Atkinson ---------- 46
        2.2.3.3   Intrinsic and extrinsic motives of Spence and Helmreich ---------- 48
        2.2.3.4   Expectancy-value model of Eccles ---------- 49
        2.2.3.5   Achievement goal framework ---------- 51

2.2.4	Some issues on achievement motivation	56
2.2.5	Motivation and achievement in collectivist and individualist cultures	57
2.2.5.1	Definitions of the two cultural syndromes	57
2.2.5.2	Collectivism	59
2.2.5.3	Implications for emotion, motivation and achievement	60
2.2.6	Hypotheses on achievement motivation	62

2.3 Affective development and academic emotions — 62

- 2.3.1 Psychological perspectives on emotion — 62
- 2.3.2 Current situation of studying emotion — 64
- 2.3.3 Definitions, components of emotion, and emotional states — 65
- 2.3.4 Formation of emotions — 67
  - 2.3.4.1 Mechanisms and phases of emotion formation — 67
  - 2.3.4.2 Brain and social development, and a cultural perspective — 68
  - 2.3.4.3 Deterrents to emotional development and individual differences in emotional development — 72
- 2.3.5 Conceptions of emotionality and regulation — 73
  - 2.3.5.1 A framework of emotionality and regulation — 73
  - 2.3.5.2 Socialization of emotional expression and regulation — 75
  - 2.3.5.3 Building emotional competence — 75
- 2.3.6 Relation with needs and values — 76
- 2.3.7 Social cognitive control-value theory of achievement emotions — 77
  - 2.3.7.1 Definition and relevance of the domain of achievement emotions — 77
  - 2.3.7.2 Assumptions of the theory — 80
- 2.3.8 Hypotheses on achievement emotions — 84

2.4 Attribution theory — 85

- 2.4.1 Entity and incremental theories of intelligence and attribution theory — 85
- 2.4.2 Development of causal structure — 86
- 2.4.3 An attribution theory of achievement motivation and emotion — 86
- 2.4.4 Motivational dynamics of perceived causality — 87
  - 2.4.4.1 Expectancy change — 87
  - 2.4.4.2 Affective reactions — 88
- 2.4.5 Chinese cultural influences on achievement attribution — 90
- 2.4.6 Achievement change programs — 91
- 2.4.7 Hypotheses on attributions — 91

2.5 Self-regulated learning — 91

- 2.5.1 Definition of self-regulated learning and learning strategies — 92
- 2.5.2 Motivational beliefs in learning — 93
  - 2.5.2.1 Expectancy component — 93
  - 2.5.2.2 Value component — 94

		2.5.2.3 Affective component	94
	2.5.3	Hypotheses on learning strategy use	95
2.6	**Translation of measures**		96
	2.6.1	Perspectives	96
	2.6.2	Procedures	97

Chapter III   Methodology — 99

3.1	**Design of the study**		99
3.2	**Sample characteristics**		99
	3.2.1	Sample size and gender	99
	3.2.2	Social characteristics of the sample	100
3.3	**Measures**		102
	3.3.1	Choosing instruments	102
	3.3.2	Reliability of the measures	105
		3.3.2.1 Motivated Strategies for Learning Questionnaire (MSLQ)	105
		3.3.2.2 Academic Emotions Questionnaire – Mathematics (AEQM)	108
		3.3.2.3 Achievement Goal Questionnaire (AGQ)	109
		3.3.2.4 Perceived Classroom Environment Questionnaire (PCEQ)	110
		3.3.2.5 Self-Referenced Cognition and Achievement values	111
		3.3.2.6 Causal Attribution Questionnaire	112
	3.3.3	Predictive validity of the measures	112
	3.3.4	Translation of the questionnaires	112
3.4	**Procedures**		118
	3.4.1	Sampling procedure	118
	3.4.2	Technical details	119
	3.4.3	Observation of the data collection process	119

Chapter IV   Results — 121

4.1	**Validity of the measures**		121
	4.1.1	Factor structure of MSLQ	121
	4.1.2	Factor structure of AEQM	122
	4.1.3	Factor structure of AGQ	123
	4.1.4	Factor structure of PCEQ	124
4.2	**Multivariate analyses of variance and covariance controlling prior math achievement**		126
	4.2.1	Attitude toward math, self, achievement values, and expectations	127
	4.2.2	Academic emotions, learning strategies, and achievement goals	136
	4.2.3	Summary of all the major variable comparisons	143

4.3	**Interrelations between affective-cognitive processes**	146
	4.3.1 Intercorrelations among variables in general	146
	4.3.2 Emotions and associations with self, values, environmental factors, and goals	148
	4.3.3 Goals and associations with self, values, and environmental factors	151
	4.3.4 Learning strategies and associations with self, values, environmental factors, emotions, and goals	153
	4.3.5 Attributions and academic emotions	155
4.4	**Mediator effect**	162
	4.4.1 Math self-concept as a mediator	162
	4.4.2 Achievement goal as a mediator	174
4.5	**Regression models**	180
	4.5.1 Regressions on academic emotions	180
	4.5.2 Regression on learning strategies	182
	4.5.3 Regression on math achievement	184
	4.5.4 Integrated picture of learning processes and achievement model	186

Chapter V Conclusions and Discussions — 193

5.1	**Factor structure of the measures among the Chinese sample**	193
	5.1.1 Learning strategies	193
	5.1.2 Achievement emotions	194
	5.1.3 Achievement goals	194
	5.1.4 Perceived classroom climate	195
5.2	**Math, a gender or achievement issue**	195
	5.2.1 A trial to understand learners' cognitive functioning	196
	5.2.2 Affective experiences in learning mathematics	197
	5.2.3 Who is more motivated to learn?	198
	5.2.4 Confidence in learning mathematics	198
	5.2.5 Parental expectancy for boys and girls	199
	5.2.6 Summary	200
5.3	**Interrelations of affective-cognitive processes**	200
	5.3.1 Associations of affective-cognitive variables	201
	5.3.2 Associations between environmental factors and affective-cognitive variables	201
5.4	**From environmental factors to emotions and from emotions to learning strategies**	202
	5.4.1 Math self-concept's mediating effect on emotions	202
	5.4.2 Achievement goals' mediating effect on learning strategies	203

5.5		**Dynamic affective-cognitive learning processes and outcome model**	204
	5.5.1	Math achievement: a multidimensional construct	204
	5.5.2	Implications for teachers, parents, and students in brief	205
5.6		**Pedagogical implications: multiple channels to support development and enhance achievement**	205
	5.6.1	Formula of development	205
	5.6.2	Enhancing emotional competence through introducing social skills	206
	5.6.3	Maintaining and enhancing self-concept	207
	5.6.4	Enhancing achievement motivation from a goal perspective	207
	5.6.5	Intervention model for improving psychological well-being and learning outcomes	210
		5.6.5.1 Intervention level I	211
		5.6.5.2 Intervention level II	212
5.7		**Limitations of the study and future research**	214
BIBLIOGRAPHY			217
BIOGRAPHY			237

List of Tables

Table 2.3.1	Classification of Achievement Emotions – Examples	78
Table 3.2.1	Descriptive Statistics of Gender, Sample Size, and School Population	99
Table 3.2.2	Socioeconomic Characteristics of the Sample	100
Table 3.2.3	SES Percentage of High and Low Achievement Samples	102
Table 3.3.1	Source Information, Item numbers, and Alpha Levels of Original Measures	105
Table 3.3.2	Scale Descriptive Statistics, Internal Reliability Coefficients, and Correlations with Math GPA for all the Measures	106
Table 3.3.3	Descriptive Statistics and t Values for the English and the Chinese Version of MSLQ	113
Table 3.3.4	A Correlation Matrix for Five Constructs of the English and the Chinese Version of MSLQ	114
Table 3.3.5	Descriptive Statistics and t Values for the English and the Chinese Version of AEQM	115
Table 3.3.6	A Correlation Matrix for Seven Constructs of the English and the Chinese Version of AEQM	115
Table 3.3.7	Descriptive Statistics and t Values for the English and the Chinese Version of AGQ	116
Table 3.3.8	A Correlation Matrix for Four Constructs of the English and the Chinese Version of AGQ	117
Table 3.3.9	Descriptive Statistics and t Values for the English and the Chinese Version of PCEQ	117
Table 3.3.10	A Correlation Matrix for Three Constructs of the English and the Chinese Version of PCEQ	118
Table 4.1.1	Factor Loadings for Achievement Goals	123
Table 4.1.2	Descriptive Statistics, Reliabilities, and Intercorrelations among Achievement Goal Variables	124
Table 4.1.3	Factor Loadings for Perceived Classroom Environment	125
Table 4.1.4	Descriptive Statistics, Reliabilities, and Intercorrelations among Perceived Classroom Environment Variables	126
Table 4.2.1	Means and Standard Deviations of Attitude toward Math, Self, Values, and Achievement Expectations of High and Low Achievement Samples	128
Table 4.2.2	Results of Multivariate Analyses of Variance and Covariance with Gender as Predictor, Mathematics Grade as Covariate, and Attitude toward Math, Self, Values, and Achievement Expectations as Dependent Variables	129
Table 4.2.3	Means and Standard Deviations of Attitude toward Math, Self, Values, and Achievement Expectations by Achievement and Gender	131
Table 4.2.4	Zero-Order Correlations of Achievement Values and Expectations between Child and Parents	132
Table 4.2.5	Tukey Test of Attitude, Self, Values, and Achievement Expectations by Achievement and Gender	134
Table 4.2.6	Means and Standard Deviations of Academic emotions, Learning Strategies and Achievement goals of High and Low achievement samples	136
Table 4.2.7	Results of Multivariate Analyses of Variance and Covariance with Gender as Predictor, Mathematics Grade as Covariate, and Emotions, Learning Strategies, and Achievement Goals as Dependent Variables	138
Table 4.2.8	Means and Standard Deviations of Achievement Emotions, Learning Strategies, and Achievement Goals by Achievement and Gender	139

Table 4.2.9	Tukey Test of Achievement Emotions, Learning Strategies, and Achievement Goals by Achievement and Gender	140
Table 4.2.10	Among Group Differences of Major Variables by Achievement and Gender	144
Table 4.3.1	Intercorrelations among Variables	146
Table 4.3.2	Descriptive Statistics of Major Variables	147
Table 4.3.3	Zero-order Correlations between Academic Emotions and Self, Values, Environmental Factors, and Achievement Goals	149
Table 4.3.4	Zero-order Correlations between Achievement Goals and Self, Values, and Environmental Factors	151
Table 4.3.5	Zero-order Correlations between Learning Strategies and Self, Values, Environmental factors, Academic Emotions, and Achievement Goals	153
Table 4.3.6	Means for Causal Attributions in Ranked Order	156
Table 4.3.7	Mean Differences and Standard Deviations of Internal Attribution and External Attribution for Success and Failure	156
Table 4.3.8	Predictors for Positive Emotions with Attributions as Independent Variables	157
Table 4.3.9	Predictors for Negative Emotions with Attributions as Independent Variables	159
Table 4.3.10	Predictors for Math GPA with Emotions and Attributions as Independent Variables	161
Table 4.4.1	Pearson Correlations among Teacher Enthusiasm, Math Self-concept, and Positive Emotions	163
Table 4.4.2	Coefficients of Teacher Enthusiasm and Math Self-concept	163
Table 4.4.3	Regression Coefficients and Standard Errors for Two Paths of Mediating Path from Teacher Enthusiasm, Math self-concept to Positive Emotions	164
Table 4.4.4	Pearson Correlations among Peer Attitude, Math Self-concept, and Positive Emotions	164
Table 4.4.5	Coefficients of Peer Attitude and Math Self-concept	165
Table 4.4.6	Regression Coefficients and Standard Errors for Two Paths of Mediating Path from Peer Attitude, Math Self-concept, to Positive Emotions	165
Table 4.4.7	Pearson Correlations among Lecture Engagement, Math Self-Concept, and Positive Emotions	166
Table 4.4.8	Coefficients of Lecture Engagement and Math Self-concept	167
Table 4.4.9	Regression Coefficients and Standard Errors for Two Paths of Mediating Path from Lecture Engagement, Math Self-concept, to Positive Emotions	167
Table 4.4.10	Pearson Correlations among Harsh Evaluation, Math Self-concept, and Negative Emotions	168
Table 4.4.11	Coefficients of Harsh Evaluation and Math Self-concept	169
Table 4.4.12	Regression Coefficients and Standard Errors for Two Paths of Mediating Path from Harsh Evaluation, Math Self-concept, to Negative Emotions	169
Table 4.4.13	Correlations among Evaluation focus, Math Self-concept, and Negative Emotions	170
Table 4.4.14	Coefficients of Evaluation Focus and Math Self-concept	170
Table 4.4.15	Regression Coefficients and Standard Errors for Two Paths of Mediating Path from Evaluation Focus, Math Self-concept, to Negative Emotions	171
Table 4.4.16	Pearson Correlations among Parental Control, Math Self-concept, and Negative Emotions	172
Table 4.4.17	Coefficients of Parental Control and Math Self-concept	172
Table 4.4.18	Regression Coefficients and Standard Errors for Two Paths of Mediating Path from Parental Control, Math Self-concept, to Negative Emotions	173
Table 4.4.19	Pearson Correlations among Positive Emotions, Approach Goal, and Learning Strategies	175

Table 4.4.20   Coefficients of Positive Emotions and Approach Achievement Goal---------------- 175
Table 4.4.21   Regression Coefficients and Standard Errors for Two Paths of Mediating Path from Positive Emotions, Approach Goals, to Learning Strategies-------------------- 175
Table 4.4.22   Pearson Correlations among Negative emotions, Approach goals, and Learning strategies-- 176
Table 4.4.23   Coefficients of Negative Emotions and Approach Goals------------------------------ 177
Table 4.4.24   Regression Coefficients and Standard Errors for Two Paths of Mediating Path from Negative Emotions, Approach Goals, to Learning Strategies------------------- 177
Table 4.4.25   Pearson Correlations among Positive Emotions, Avoidance Goals, and Learning Strategies-- 178
Table 4.4.26   Pearson Correlations of Avoidance Goals, and Learning Strategies------------------ 179
Table 4.5.1   Pearson Correlations of Emotions and Potential Predictors--------------------------- 180
Table 4.5.2   Multiple Regression Analysis for Variables Predicting Positive Academic Emotions-- 181
Table 4.5.3   Multiple Regression Analysis for Variables Predicting Negative Academic Emotions-- 182
Table 4.5.4   Multiple Regression Analysis for Variables Predicting Learning Strategy Use---- 183
Table 4.5.5   Multiple Regression Analysis for Variables Predicting Math Achievement-------- 185
Table 4.5.6   Summary of Predictors for Academic Emotions, Learning Strategies, and Math GPA--- 187

List of Figures

Figure 1.5.1 School System in China ---------- 33
Figure 2.2.1 General Expectancy-Value and Developmental Model of Achievement Behaviors (Eccles, 1983) ---------- 52
Figure 2.2.2 The 2 Times 2 Achievement Goal Framework (Elliot &McGregor, 2001) ---------- 55
Figure 2.3.1 Emotional States (Bruno, 1992) ---------- 67
Figure 2.3.2 Illustration of Bowlby's (1973) Developmental Pathways Concept (Sroufe, L. A., 1996) ---------- 70
Figure 2.3.3 Neo-Piagetian Theory of Emotional Development (Adapted from Piaget, 1967) --- 76
Figure 2.3.4 Representation of Goals as the Transactive Point Among Cognition, Motivation, and Emotion (Schutz et al., 2006) ---------- 77
Figure 2.3.5 Social Cognitive Control-Value Theory Model (adapted from Pekrun, 2000) ---------- 83
Figure 2.4.1 The Cognition-Emotions Process (Weiner, 1985) ---------- 88
Figure 4.2.1 Self Achievement Expectation and Achievement Expectation from Parents of High and Low Achievement Samples ---------- 128
Figure 4.2.2 Difference Lines of Self Achievement Expectation and Achievement Expectation from Parents for Female and Male Samples ---------- 130
Figure 4.2.3 Self and Attitude towards Mathematics by Achievement and Gender ---------- 133
Figure 4.2.4 Achievement Expectations by Achievement and Gender ---------- 133
Figure 4.5.1 A Summary of the Results from Joint Regression Analyses ---------- 189
Figure 4.5.2 Stepwise Regression for Math Achievement ---------- 190
Figure 5.6.1 Schematic Representation of Goal Theory Model (Anderman and Maehr, 1994) -- 210

List of Diagrams

Diagram 4.4.1 Math Self-concept as a Mediator on Achievement Emotions------------------- 174
Diagram 4.4.2 Approach Goal as a Mediator on Learning Strategies--------------------------- 179
Diagram 4.5.1 Affective-Cognitive Learning Processes and Achievement Model------------ 191
Diagram 5.6.1 Intervention Model for Improving Psychological Well-Being and Learning Outcome-- 213

Note:

Each table, figure and diagram is represented by three code numbers. The first code is the chapter where a table, figure, or diagram is situated; the second code is the subchapter where a table, figure, or diagram is found exactly; the third code is the order of a table, figure, or diagram in the subchapter. For example, Table 4.1.2 Descriptive Statistics, Reliabilities, and Intercorrelations among Achievement Goal Variables, is the second table in the first subchapter (Validity of the measures) of Chapter IV Results.

Chapter I INTRODUCTION

1.1 Purpose of the study

The purpose of the study was to understand adolescents' affective, motivational, and cognitive processes in learning mathematics. Understanding the interrelations among students' cognitive, emotional, and motivational processes is an emergening focus in educational psychology (Eynde & Turner, 2006). Affective, motivational, and cognitive processes are interdependent in the learning-related experiences of students. At the mean time they are conceptually and empirically separable. Affect is as central to understanding the character of educational experiences as are motivation and cognition (Ainley, 2006).

The focus was adolescents and their affective, motivational, and cognitive processes in learning mathematics. The differences of affective-cognitive processes were investigated between males and females, high achievers and low achievers, and gender by achievement groups, respectively. Six most frequently studied affective phenomena: emotions, feelings, moods, attitudes, affective styles, and temperaments were identified in the introduction to the Handbook of Affective Sciences (Davidson, Scherer, & Goldsmith, 2003). The affective processes in the current study referred to academic emotions, achievement goal orientations, control and value beliefs, attributions, expectancies towards learning and achievement, etc., with academic emotions being the focal point. The cognitive processes mainly referred to learning strategies in learning mathematics. Environmental factors' impact on academic emotions, and achievement goals were investigated as well.

Academic emotions in learning mathematics in China and Germany were investigated (Frenzel, Thrash, Pekrun, Goetz, 2007c; He, 2004; Wan, 2004; Ye, 2004) from social cognitive control-value perspective, which was also one of the theoretical foundations for the current study. The comparisons of academic emotions within cultures of independent and interdependent self-construes were the focus of He (2004) and Ye's (2004) studies. They did not compare academic emotions between gender groups, achievement groups, or gender by achievement groups. Differences in mean levels of achievement emotions were found across cultures. Chinese students were found to experience more enjoyment, pride, shame, anxiety and less anger than German students (Frenzel et al., 2007c; He, 2004; Wan, 2004; Ye, 2004). Construct comparability of achievement emotions across cultures was established in Frenzel et al.'s study (2007c).

The current study was based on broadened perspectives and integrative approaches of affect, motivation, and cognition in educational context and took a closer look at the constructs of affect, motivation, and cognition in a collectivist culture. The current study also aimed at designing appropriate and practical pedagogical interventions for teachers, parents, and students. Specifically, the issues, e.g., person-environment fit of motivation and instruction and its potential pedagogical implication for motivational scaffolding, use of learning strategies and its pedagogical implication for cognitive scaffolding, emotion regulation and emotion competence, etc., were addressed for the purpose of intervention in the educational context of learning mathematics.

1.1.1 Adolescence – a critical point

With its origins in the Latin word *adolescere*, meaning "to grow into adulthood," the period is initiated by the onset of puberty and comes to a close with the full transition to adult status (LaFreniere, 2000). The transition from middle childhood to adolescence is marked by a number of significant biological and cognitive changes. The final task of adolescence is to complete the integration of the developmental changes (biological, social, psychological, cognitive) into a well-adjusted young adult (Jones, 1992).

This term was first used by the Belgian anthropologist Arnold van Gennep (1960), who noted that traditional societies practiced a three-part transitionary ritual involving, an initial isolation phase, a phase of intentional confusion in which the identity of the adolescent is broken down, and a phase of reinstatement in which the adolescent is incorporated into the adult community. Human ethnologists also suggest that puberty rites function to provide instruction in adult sex roles and to instill cultural loyalty (Weisfeld, 1997, 1999). The American psychologist G. Stanley Hall is generally credited for the first comprehensive scientific treatment of adolescence (Hall, 1904). Influenced by Darwin, Hall characterized adolescence as a period of "storm and stress," a term that he chose to signify the passionate idealism and commitment to revolutionary principles that embodied the late eighteenth-century German literary and political movement known as "Strum und Drang." Although many contemporary American psychologists believe that the "Strurm und Drang" of adolescence has been greatly exaggerated (Steinberg, 1990), Hall's way of looking at adolescence has prevailed. On the whole, it is thought of as a period of unusual emotional turbulence (Bruno, 1992, p.6).

Biosocial perspective on adolescence
- *Biological changes*

From the point of view of biological development, a span of time starts with puberty and concludes with maturity. Chronologically, this is usually from about 12 or 13 to 18 or 19 years of age. (Consequently, there are the informal terms *teenage years* and *teenager.)* Some authorities suggest that adolescence ends for females at about the age of 21 and for males at about the age of 22 (Bruno, 1992, p. 6). Due to the activating effects of hormones, puberty dramatic morphological changes occur. These changes include growth spurts, qualitative changes in physical appearance, the emergence of primary and secondary sex characteristics, and a general increase in sexual dimorphism (LaFreniere, 2000, p.242).

- *Cognitive changes*

From the view of cognitive development, the adolescent becomes capable of formal operations, which enable them to deal with relatively complex mathematics and math related subjects. Cognitive changes in the adolescent include the acquisition of abstract thought processes, logical reasoning, hypothesizing, and metacognitive thinking skills. These higher-level cognitive abilities provide the adolescent the ability to engage in complex perspective-taking in fundamental issues such as social relations, morality, politics, and religion (Jones, 1992b).

- *Psychological changes*

During adolescence emotions noticeably become the basis of identity and ideals. Adolescents begin to form varied and intense attachments to ideals, people, and careers, just as their emotional life is transformed and as awareness of emotionality changes. Becoming aware of emotionality changes is one of the psychological changes happening during adolescence. Social context of emotions is shifted away from a family focus toward a peer focus. This pattern of change is consistent with adolescents' general social development (Haviland, Gebelt, & Stapley, 1997).

Awareness of emotion One of Stapley and Haviland's (1989) studies gathered information about fifth-, seventh-, and ninth-graders' awareness of their emotions. They found that it was not until high school, well past the hormonal surges of puberty, that students were likely to report mood changes. Trying to abstract the meaning of moods and personality traits from concrete experience was difficult for adolescents.

The easiest emotions to accept and comprehend are the positive emotions (joy, surprise, and interest). These are reported by all of the students in all the grades to be much more changeable than other emotions. Even college students prefer to think of themselves in terms of degrees of happiness rather than degrees of sadness, anger, or fear (Gebelt, 1995). Young adolescents have trouble acknowledging and tracking their negative emotions (Stapley and Haviland, 1989).

It is a common psychological reaction at all ages to deny a change occurs. It is only after the changes are partially assimilated that some awareness of them is gained. The assimilation of negative emotions, e.g., sadness, anger, and fear, etc., is especially slow. Emotional refinement includes increasing awareness of mood changes, especially internally directed ones. It also includes changes in the perceived organization of social situations that elicit emotions. They transform the social and personal scenes of the child to those of the adolescent. Getting in touch with one's feeling and emotions is a sign of maturity (Haviland, Gebelt, & Stapley, 1997). Being aware of one's feelings and emotions is helpful for emotion regulation.

Elicitors of emotion Analyses of the adolescents' responses suggested that there are systematic developmental differences for the social context of emotion. The patterns are consistent with general social development away from a family focus and toward a peer focus. The older the student, the less likely family members are to be mentioned, and this will be particularly obvious with respect to positive emotions. Emotions are being attached to friends, especially romantic ones, and situations in which friends are the key participants. In contrast, for children the key people are members of the family and the situations are family scenes (Haviland, Gebelt, & Stapley, 1997, p.246).

- *Social changes*

Socially during middle childhood, the child learns to function in three separate worlds – the family, the school, and the peer group – each with its own rules and styles of behavior. Major social interactions during adolescence center around peers and friends. Peer groups serve as a socializing agent and friends provide the support needed during this period of self-concept building (Jones, 1992).

- *Summary*

According to Hill's framework (Hill, 1980), adolescent development is set into motion by the primary changes in biology, cognition, and social relations and the secondary psychosocial changes, namely, attachment, autonomy, sexuality, intimacy, achievement, and identity. Primary changes are thought to be universal aspects of the adolescent experience. Concerning the secondary changes, adolescent change in attachment and autonomy mainly display itself in family relation transformation; adolescent change in sexuality and intimacy display itself mainly in peer relation change; adolescent change in achievement and identity mainly display itself in transformation of the self.

Developmental tasks for adolescents

The developmental changes: biological, social, psychological, cognitive changes can be reflected in the following tasks.

- *Mature socially and physically*

The average adolescent understands and accepts the biological changes happening to their bodies. Social maturation requires that adolescents understand the parameters of maleness and femaleness, become cognizant of adult responsibilities associated with their sexuality (Jones, 1992).

- *Social interactions: peer group and friendships*

School is the major arena in which adolescents earn status. Good peer relationships appear to be necessary for normal social development in adolescence (Jones, 1992). Friendships serve a number of important functions including the companionship of a familiar partner, source of interesting information, physical support and assistance, ego support in terms of encouragement, social comparison, and an intimate trusting relationship (Santrock, 1990).

- *Adolescent self-concept*

For the adolescent, self-concept development involves constructing a self-portrait which integrates the different characteristics of the self. This process includes realistically evaluating his/her intellectual competence, physical competence, physical attractiveness, social competence, leadership abilities, moral beliefs, and sense of humor (Jones, 1992). For the adolescent, the self-concept now functions as a standard for evaluating and predicting performance socially, objectively, personally; and functions to limit performance for the purpose of maintaining and enhancing itself (Manaster, 1989). "Compared to younger children, early adolescents are highly self-conscious and have uncertain, shaky images themselves with regard to certain qualities they value" (Skolnick, 1986, 463). Adolescents are able to maintain a reasonably positive self-image by identifying positive attributes as core constructs in their self-portrait and considering negative attributes or behaviors as foreign to their true self (Harter, 1986).

- *Making career/vocational choices*

A primary task of adolescence is to select a career or vocation to pursue. The adolescent needs to be cognizant of the numerous social factors which impact this decision including the following: (1) socio-economic status of the family, (2) the education level of the parents, (3) the school-community environment, (4) the degree of the adolescent's status seeking among peers, (5) the adolescent's own personal aptitude and achievement, and (6) the adolescent's desire for economic independence (Jones, 1992b). Functioning as a responsible citizen and desatillizing from home are also challenging tasks for adolescents (Jones, 1992).

1.1.2 Learning mathematics – cognitive functioning with affect

Biological and psychological changes make adolescence a critical point in life for adolescents. Mathematics is one of the core courses which an adolescent has to deal with in his or her daily academic life. It prepares an adolescent for more advanced career oriented training and consequently choosing his or her career in the future. The complexity of mathematic problems overwhelms some adolescents from time to time and many more others frequently. Therefore, it is necessary to understand how adolescents deal with their emotions in math learning and whether and how emotions are related to achievement motivation and cognitive functioning. It is also interesting to know whether motivation and self-regulated learning components operate independently or jointly to influence student academic performance in the classroom. In addition, it represents a unique opportunity to explore gender differences in an area in which girls are stereotypically expected to do worse than boys.

1.2 Collectivist and individualist cultures and implications for motivation

Traditional (western) theories, suggest that the achievement motivation trait is developed through childrearing processes which emphasize independence, mastery and competitiveness (McClelland, Atkinson, Clark, & Lowell, 1953). Cross-cultural researchers in particular have reported that in many cultures achievement behavior does not confirm with the western prototype described by McClelland and his associates (cf. De Vos, 1968; Maehr, 1978), It has been suggested that achievement should be understood and defined in terms of the social and cultural context of individuals (Salli, 1995).

Individualism and collectivism are cultural syndromes. A cultural syndrome consists of shared attitudes, beliefs, norms, roles, values, and other elements of subjective culture (Triandis, 1972), found among those who share a language, historical period, and geographic locations (Salili, 1995). Achievement for individualists is individual achievement, and is often seen as a means for self-glory, fame, and immortality (Salili, 1995). Individualists focus on the achievement of personal goals, by themselves, for the purpose of pleasure, autonomy, and self-realization. Collectivists focus on the achievement of group goals, by the group, for the purpose of group well-being, relationships, togetherness, the common good, and collective utility (Triandis, 1995).

Central to Chinese cultural values is Confucian teaching and philosophy, which is the corner stone of Chinese value system and collectivism. Chinese attach great importance to filial piety which emphasizes respect and unquestioning obedience and obligations towards one's parents (Hofstede, 1983). Influenced by Confucian philosophy, the Chinese place great emphasis on education. Achievement through effort, hard work and endurance is highly emphasized and working hard and achieving beyond one's abilities is considered a virtue (Yang, 1986).

Collectivism has important implications for motivation. Wilson and Pusey (1982) provided evidence that the Chinese place greater importance on family and group goals than on individual goals in achievement and they are very concerned about loss or gain of their collective face in their pursuit of achievement. Achievement motivation among the Chinese is mediated more by affiliative concerns than personal needs and that their reasons for achievement and meaning of achievement may be different from that of western nations (Salili, 1995).

Theory and empirical research show that there are differences in mean levels of achievement motivation and other achievement-related constructs across Eastern and Western countries (e.g., Chen & Stevenson, 1995; Frenzel et al., 2007c; Salili, 1994). In a cross-cultural comparison of German and Chinese middle school students' achievement emotions, structurally equivalent relations of appraisals and emotions were found. However, mean levels of emotions differed between cultures, with Chinese students reporting significantly more achievement-related enjoyment, pride, anxiety, and shame, and significantly less anger than German students (Frenzel et al., 2007c).

1.3 Cognitive and motivational characteristics of different groups in math learning context

1.3.1 High versus low achievement groups

The researcher was also interested to find out the cognitive and motivational characteristics of different levels of ability groups, namely academically gifted adolescents (high achievement group) and their nongifted peers (moderate and low achievement groups). Research suggests that math and science achievement are positively associated with youths' values and self-concepts (e.g., Casey, Nuttall, & Pezaris, 1997; Frome & Eccles, 1998; Jacobs, Finken, Griffin, &Wright, 1998; Updegraff, Eccles, Barber, O'Brien, 1996). Cognitive and motivational characteristics of the academically gifted have been examined by comparing them with those of their nongifted peers. Academically gifted children are cognitively more competent and are more intrinsically motivated than their nongifted peers (Gottfried & Gottfried, 1996; Vallerand, Gagné, Senécal, & Pelletier, 1994). Gifted students tend to be more strategic (Montague & Applegate, 1993; Shore & Dover, 1987), use learning strategies more effectively and transfer those strategies to novel tasks when trained (Riesemberg & Zimmerman, 1992), and use more rereading, inferring, analyzing structure, predicting, and evaluating strategies (Fehrenbach, 1991) when compared to their nongifted counterparts.

1.3.2 Male versus female groups

Cognitive sex differences and mathematics and science achievement

Gender differences are a pervasive theme throughout the literature on math and science. Math and physical science are often considered to be domains in which boys have high achievement, values, and self-concepts (Simpkins, Davis-Kean, and Eccles, 2006). Boys are more likely than girls to enroll in math courses during high school (e.g., Farmer, Wardrop, Anderson, & Risinger, 1995; Updegraff et al., 1996). However, in her critical review of cognitive sex differences in mathematics and science achievement, Spelke claimed (2005, p.954) that "most investigators of sex differences have concluded that males and females have equal cognitive abilities, with somewhat different profiles." She (p.956) further concluded that "research on the cognitive abilities of males and females, from birth to maturity, does not support the claim that men have greater intrinsic aptitude for mathematics and science" and suggested that "we look beyond cognitive ability and to other aspects of human biology and society for insights into the phenomenon" of preponderance of men on academic faculties of mathematics and science. For some abilities, the differences between sex groups is small (e.g., as in the case of abstract reasoning – see Lynn & Irwing, 2005), but for other abilities and differences are quite large (e.g., mathematical reasoning – see Halpern, 2000). Ackerman (2006, p.722) therefore stated that "whether males or females have higher mean general intelligence depends on the operationlization of the content of the tests selected to assess cognitive ability." According to Ackerman (2006), early test developers dealt with this problem in two strikingly different ways. Yerkes, Bridges, and Hardwick (1915) required separate norms for males and females to deal with these differences, whereas Terman and Merrill (1937, p.34) ensured equal mean levels for boys and girls by eliminating individual scores in which large gender differences were found.

Geary (1995) pointed out that mathematical learning for the most part involves biologically secondary abilities, which build on biologically primary abilities yet can only be nurtured through cultural provisions. Geary's (1995) research showed that girls' mathematical reasoning can be improved by telling them to use spatial strategy. Dai (2006) commented that aptitude is likely developmental in nature, subject to both genetic and environmental influences and likely reflects a complex interplay of nature and nurture.

In the report of Advanced Placement testing statistics (College Board, 2005), the gender differences on math and science performance are compelling, in terms of numbers of students taking the tests and the number of "passing" scores. Although females completed 155,849 more AP tests across all content areas than males, more males completed the Calculus AB test (7,439) than females, and a higher percentage of males (43.8%) than female (35.5%) had clear passing scores (of 4 or 5). Spelke (2005) stated that sex differences in basic cognitive abilities do not explain "the differential representation of men and women in high-level careers in mathematics and science" (p.950). Domain knowledge (Ackerman, 2006) provided evidence why there is such a differential representation.

Motivational and affective sex differences and math achievement

Dai (2006) claimed that it is meaningless to focus only on the question projected by Spelke (2005) "Do men and women have equal cognitive capacities for math and science careers?". Spelke's (2005) review found no differences either in the core mathematical abilities of male and female infants or in the developed mathematical mastery of college men and women. These findings are consistent with the two following conclusions. First, the motivations and strategies that interact with mathematical development may be shared by boys and girls. Second, the development of mathematical thinking may be robust over variations in motivation and strategies (Spelke & Grace, 2006, p.725).

According to Dai (2006, p.723), Spelke's question quoted above represents "a narrow, performance-on-demand view of aptitude, as aptitude concerns not only what one can do when given a task (i.e., capacity) but also what one will do (i.e., conation) and how one will do (i.e., strategy development and style) given a situation". A more dynamic, contextual conception of aptitude is theoretically more viable because molarlevel intellectual functioning in real-life contexts is never a mechanical switching on and off of some invariable performance-on-demand capacity, but rather a dynamic interplay of cognition (both automatic and controlled processes), affect, and conation (Snow, Corno, & Jackson, 1996). Affect and conation regulate attention and cognition not only quantitatively but also qualitatively (Dweck, Mangels, & Good, 2004), and they sometimes transform cognition (Dai, 2006).

The investigator (Eccles, 1983) tested the interrelationships among the cognitive factors specified and the contribution of these psychological variables to a measure with implications for actual achievement behavior: the students' intention to take additional math courses. These analyses confirmed the importance of children's self-concepts of ability, attributions for past performance, and perceptions of the beliefs of parents and teachers as determinants of expectancies, values, and course plans. Relatively few sex differences were found, but those that did appear confirmed the results of past investigations: females, in comparison with males, had lower confidence in their ability and perceived math as more difficult. Such sex differences appeared to be related to parents' beliefs in the difficulty of math for their child. Similar results further confirmed that girls demonstrate less self-confidence in their mathematical ability than do boys quite early (Iben, 1991; Tiedemann, 2000) and girls have lower math and science self-concept than boys (e.g., Andre, Whigham, Hendrickson, & Chambers, 1999; Jacobs, Lanza, Osgood, Eccles, & Wigfield, 2002; Updegraff et al., 1996). Findings on youths' overall value, which includes both interest and beliefs about importance, suggest that boys and girls report similar values of mathematics and science (Andre et al., 1999; Eccles et al., 1993c; Jacobs et al., 2002; Simpkins et al., 2006; Wigfield, Eccles, Mac Iver, Reuman, & Midgley, 1991).

Emotions depend on control and value appraisals in both girls and boys. However, to the extent that perceived control and academic values differ between girls and boys, resulting emotional experiences can differ (Pekrun, 2006). In recent studies (Frenzel, Pekrun, Goetz, & vom Hofe, 2006; Frenzel, Pekrun, & Goetz, 2007b) on gender differences in upper elementary students' achievement emotions in mathematics, relations between appraisals and emotions were structurally equivalent across the genders. However, mean scores for perceived control were significantly lower in girls and these differences in emotions proved to be mediated by the gender differences of appraisals. Frenzel et al. (2007b) found that girls reported significantly less enjoyment and pride than boys, but more anxiety, hopelessness, and shame. A meta-analysis by Hyde, Fennema, Ryan, Frost, and Hopp (1990) showed that gender-based discrepancies of affect in mathematics are prevalent and of considerable size. Gender discrepancies in emotional experiences were found even though girls and boys had achieved at similar levels in mathematics, as evidenced by their mid-term grades (OECD, 2004). Gender-linked stereotypes of domain-related abilities are a possible reason for explaining the discrepancy between gender differences found for achievement and those for emotions (Frenzel et al., 2007b).

Stereotype threat and math achievement

The stereotype threat model postulates a situational decrement in a person's performance due to the awareness that his or her own ingroup is rated as less skillful in the domain in which he or she is going to be tested (Steele, Spencer, & Aronson, 2002). Girls may experience greater anxiety than boys because they feel that they will be judged by the cultural stereotypes about girls' lack of ability in math. This heightened anxiety interferes with girls' ability to produce effective strategies and ultimately inhibits performance (e.g., Quinn & Spencer, 2001). Research suggests that the stereotype threat model accounts for at least part of female underperformance in mathematics tests (e.g., Davies, Spencer, Quinn, & Gerhardstein, 2002; Shih, Pittinsky, & Ambady, 1999). Muzzatti and Agnoli (2007) further found that the process of gender stereotyping of mathematics began in Grade Two and got fully established in Grade Five in primary school but disappeared in Grade Eight in middle school.

Eccles and Jacob (1986) claimed that gender differences on standardized math tests were mostly rooted in stereotypic gender-role beliefs that parents (and teachers) communicate to their children (and students) on a daily basis, resulting in differential expectations, confidence, and attitudes toward math in boys and girls (see also Eccles, Adler, & Meece, 1984; Eccles, Jacob, & Harold, 1990; Jacobs & Eccles, 1992). Consistent with this, Eccles et al. (1990) showed that parents' beliefs about their daughters' versus sons' math abilities were related subsequently to their child's math self-efficacy, identification with math, and math performance.

1.4 Research questions

The critical point of adolescence and the complex nature of cognitive and affective involvement in math learning besides math being regarded as a masculine domain made the following four research questions be the focal point of the study:

I   What are the dynamics of adolescents' cognition-emotion processes, namely the interrelations between self-concept, achievement values, emotions and motivation, in the achievement-related context of math learning?

II   How do social and individual antecedents associate with adolescents' academic emotions and achievement motivation?

III   Do high achievers' cognition-emotion processes differ from low achievers' processes? Do male students' cognition-emotion processes differ from female students' processes?

IV   How does learning environment influence students' affective experiences and how do students' affective experiences influence learning strategy use?

1.5 School system in China

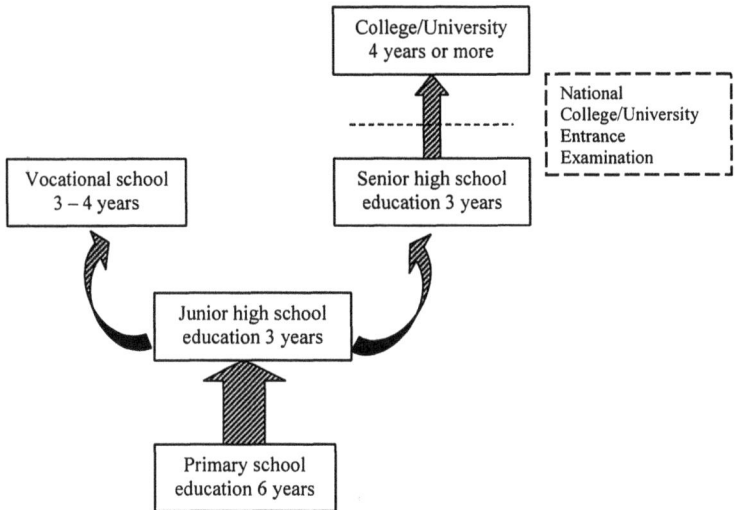

Figure 1.5.1: School System in China

The compulsory school education includes six years of primary school education and three years of junior high school education for students between the ages of 6 and 15. After receiving the compulsory education students can choose either vocational schools which prepare them for their vocational life in three years' time or senior high schools which prepare them for the National College/University Entrance Examination in three years' time. Students who have competitive scores in the National College/University Entrance Examination can choose elite tertiary institutions, e.g., first class universities. Students who have moderate scores can choose ordinary tertiary institutions. Students who have low scores have limited opportunities to go to university.

The context of learning in Hong Kong and other Chinese communities has been characterized as highly structured and authoritarian with large class sizes, expository teaching method and excessive homework (Biggs, 1992). Students listen and take notes of the teacher's lectures which cover what the examination syllabus requires. Teaching and learning activities are focused on preparing the students to pass endless tests and examinations (Salili, 1995). Students have to go through many competitive examinations in their school years. Most schools have at least two formal examinations every year and informal tests at regular intervals (e.g., monthly).

The standard set for the curriculum is very high especially that of high school curriculum. Many children of average ability need help in the form of private tutoring to pass their exams. Many students who are unable to cope with the extraordinary demands and pressure from school work and examinations eventually drop out of school. Therefore, a high level of achievement motivation is needed to survive in the education system.

Chapter II   THEORETICAL BACKGROUND

2.1   Self-esteem and self-concept

2.1.1 Self-esteem

Definition

The term self-esteem is now becoming more universally understood. It is commonly used to refer to the evaluations people make and maintain of themselves. It includes "attitudes of approval or disapproval and the degree to which people feel worthy, capable, significant, and effective" (Reasoner, 1992, p.2). Persons with a negative self-concept tend to have low self-esteem. Persons with a positive self-concept tend to have high self-esteem (Bruno, 1992).

William James (1890), perhaps the founder of self-esteem psychology, saw self-esteem as the discrepancy between one's ideal self and one's perceived self. Briggs (1970) reported that self-esteem is the sum of one's feelings about oneself, including the sense of self-respect and self-worth. These feelings, she stated, are based on the conviction that the person is a) lovable and b) worthwhile – meaning that individuals are competent enough to handle themselves and their environment and have something to offer others.

Bean (1992) equates self-esteem with the feeling of satisfaction that arises when individual needs are met. This occurs by the way people handle the world or are able to influence events through their own abilities and by how they are influenced by the world or their environment. William Purkey (1990) states that self-esteem is a habit of seeing oneself in a given way and one's disposition to expect himself or herself to succeed, to be competent, sufficient, lovable and capable in any future situation.

The most widely accepted definition of self-esteem today is one that was initially developed by Nathanial Branden (1993) and adopted by the National Council for Self-esteem, which defines self-esteem as "the experience of being capable of managing life's challenges and feeling worthy of happiness." Self-esteem involves both a cognitive and an affective

process. It develops from the cognitive process of evaluating oneself and the affective process of feeling one' worth in six areas:
(1) Inherited endowments, such as intelligence, appearance, and natural abilities
(2) Moral virtue or integrity
(3) Accomplishments or successes in life, such as skills, possessions, and achievements
(4) Feeling likeable and lovable
(5) Feeling unique, of value, and worthy of respect
(6) Feeling in control of one's life

According to Reasoner, self-esteem is not typically arrived at through competition with others, nor is it a tally of one's successes and failures. It is based more on a feeling of competence or efficacy in dealing with the future than it is on satisfaction with past accomplishments. Self-esteem is used to denote the sum total of the feeling one has about one's multiple self-concepts or self-images (1992, p.3).

Formation

Stanley Coopersmith (1967) studied the home conditions of 1,730 families to determine what patterns of adult behavior significantly altered the level of children's self-esteem. He identified three basic conditions that contributed to high self-esteem in the home environment: (1) unconditional love, (2) well-defined limits consistently enforced, and (3) a clear amount of respect shown to children. On the whole, parents who tend to be democratic, authoritative, and affectionate foster self-esteem in their children. As an adult, a series of failures in important life tasks can undermine self-esteem. Conversely, a series of successes can bolster self-esteem (Bruno, 1992, p. 329).

- *How self-esteem develops*

Children reportedly begin to form their first feelings of self-esteem as early as six weeks, based on their assessment of how the world responds to their physical and emotional needs. As children go through various stages of development, their self-esteem is modified in relation to how the important adults in their lives respond to their needs and the degree of success they have in negotiating through each developmental stage (Reasoner, 1992, p.4). According to Reasoner (1992, pp. 4-5), these are the critical stages when self-esteem develops:

Infancy stage The formation of attitudes starts to develop as children begin to reach out to others, make demands, and begin to walk. When these efforts are met with positive responses and encouragement, children begin to develop a sense of confidence. Otherwise, children will feel unwanted or less important.

Kindergarten stage By the time children enter kindergarten, the basic elements of self-esteem are in place. Children with high self-esteem mix in easily with other children, enjoy new experiences, are curious and ask questions, and respond to challenges. Children with low self-esteem are apt to have a difficult time separating from their parents, keep distant and enter activities only when it appears safe, rarely ask questions or volunteer answers.

Middle childhood With the school years come increased self-awareness, and self-consciousness. Elementary school children equate their feeling of importance with the amount of attention they receive. Teachers can have significant impact on the self-esteem of their students.

Adolescence stage As children approach adolescence, the factors that have the most influence on their self-esteem begin to shift from pleasing adults to gaining peer acceptance. As students enter high school, test anxiety increases and the significance of social popularity decreases. More significance is attached to one's assessment of ability.

Adulthood stage With the onset of adulthood, greater significance is placed on success, achievement, and income, while less significance is attached to the factors that were so important in adolescence – athletic prowess and popularity.

- *Portraits of students with high and low self-esteem*

According to Reasoner (1992, pp.5-6), students with high self-esteem view themselves realistically and accept themselves as being okay. They can identify their strengths and acknowledge their limitations. In addition, they set goals for themselves. Students with low self-esteem seem more concerned with preserving their sense of self-respect or "failing with honor" than with putting forth the extra effort needed to succeed. They typically engage in defensive behaviors to prevent others from knowing how insecure and inadequate they feel. These defensive mechanisms might include (1) failing to take responsibility for their actions, (2) withdrawing, being shy, or daydreaming, (3) blaming others when things don't go right, etc..

Stanley Coopersmith (1967) studied fifth- and sixth-grade children's self-esteem. A part of the research dealing with boys showed that those who had a high degree of self-esteem were active, expressive children who were successful both academically and socially. Boys in the middle ranges of self-esteem were quite similar to the high-self-esteem boys, but with more conventional values and behavior patterns. The low-self-esteem boys tended to be discouraged, depressed, and convinced of their inferiority. Another surprising finding was that the parents of high-self-esteem children were stricter and less permissive than those with medium and low self-esteem. They exerted greater demands than the other parents for academic performance and excellence and they were less punitive than other parents in the survey.

Components

Reasoner (1992) termed the five components of self-esteem as: sense of security, sense of identity, sense of belonging, sense of purpose, and sense of competence. The five components of self esteem are derived from Robert Reasoner's (1982) exhaustive review of self theory. As the child experiences each component he or she is involved in self evaluation. This process of evaluating or judging inner self descriptions is self esteem and is manifest in how children feel about themselves from within. According to Robert Reasoner (1982), the five components are:

Security Security is a feeling of strong assuredness that involves: feeling comfortable and safe, knowing what is expected, being able to depend on individuals and situations and comprehending rules and limits. A solid feeling of security (I can count on others) is the foundation upon which the other components within the model are built. Reasoner (1994) claimed that the sense of security is a basic human need.

Selfhood / Self-concept Reasoner (1992) reframed this selfhood as identity, which is referred to as self-concept or self-image and is seen as an element of self-esteem. A sense of identity is used to connote the sense of personal awareness or self-perceptions individuals have about themselves in terms of roles, attributes, and physical characteristics. Identity or self-concept might be described as the picture one holds of oneself; self-esteem is the feeling one has about that picture (p.35). William Purkey (1978) states that identity, or the self-concept, doesn't determine behavior, but rather guides the individual in choosing behaviors consistent with that identity.

Affiliation Affiliation is a feeling of acceptance, or relatedness, particularly in relationships that are considered important. It also means feeling approved of, appreciated and respected by others. Inner conflict can arise when the need for achievement and excellence and the need for social acceptance clash (Lofgreen & Larson, 1992). In studies of gifted adolescents, (Jenkins-Friedman & Murphy, 1988; Kerr, Colangelo & Gaeth, 1988; Robinson, 1990) students were both pleased with their capacity to perform in talent areas and concerned about possible negative perceptions of peers.

Mission Mission is a feeling of purpose and motivation in life. It allows for self empowerment through setting realistic and achievable goals and being willing to take responsibility for the consequences of one's decisions.

Competence Competence is a feeling of success and accomplishment in things regarded as important or valuable. It is being aware of strengths and the ability to accept weaknesses.

2.1.2 Self-concept

2.1.2.1 Origins and development of self-concept

- *Origins of self-concept*

Reasoner regarded self-concept or self-identity as one of the elements of self-esteem. According to Reasoner (1992), a sense of identity is used to connote the sense of personal awareness or self-perceptions individuals have about themselves. Identity or self-concept might be described as the picture one holds of oneself; self-esteem is the feeling one has about that picture (p.35).

According to Bruno (1992, p. 328), self-concept is a set of related ideas that one holds about oneself in terms of intelligence, creativity, interests, aptitudes, behavioral traits, and personal appearance. The self-concept may be generally positive or generally negative. If it is generally positive, then the individual usually feels able and attractive. If it is generally negative, then the individual usually feels inadequate and unattractive. Self-concept refers to self-perceptions formed through experience with the environment and, in particular, through environmental reinforcements and the reflected appraisals of others (Marsh & Craven, 1997) and is typically measured at a higher generality than self-efficacy (Pajares & Miller, 1994). Self-efficacy refers to beliefs in one's capabilities to organize and execute courses of action required to achieve certain performance outcomes (Bandura, 1997).

To a highly significant degree, the self is a social product. The self-concept, as Harry Stack Sullivan (1947) once said, is composed of the "reflected appraisals" of others. Through learning the opinions, attitudes, and expectations that others have for him, the child learns "who he is." It is therefore understandable how either exaggerated praise or chronic belittling can have unfortunate effects on the child's self-concept or self-image (Watson & Lindgren, 1979, p.334).

- *Development of self-concept*

The way a child is viewed differs from person to person even in a short space of time. How the child integrates all these views of himself and develops a concept of "who he is" is a highly individualized process, based not only on the views of others, but also on his own temperament and experiences, past and present (Watson & Lindgren, 1979, p.335).

Infancy stage The development of self-concept begins in infancy with self-awareness and developing a body image. Moving out into the world leads the infant to an awareness of separateness from the caregiver – a necessary but temporarily deflating step in the formation of self-concept (Breger, 1974).

Toddler period During the toddler period, developments in cognition, social behavior, and physical maturation converge to issue in both new emotional experiences and new capacities for regulating emotion, and these new experiences and capacities, in turn, promote cognitive growth, including advances in self-awareness (Sroufe, 1996, pp.193).

Primary grades to middle childhood The major achievement during the primary grades is to develop a global self-concept. By late middle childhood, the child has become efficient in using social perspective-taking, social competence, self-evaluation, self-regulation, and social problem-solving.

Adolescence stage Self-concept development during adolescence involves constructing a self-portrait in which the different characteristics of the self are integrated. This process includes realistically evaluating his/her intellectual competence, physical competence, physical attractiveness, social competence, leadership abilities, moral beliefs, and sense of humor. The adolescent develops the ability to use the perspective of a universal person, and to use the self-concept as a standard for evaluating and predicting performance and enhancing itself (Jones, 1992).

2.1.2.2 Subjective and objective selves

Cooley and James proposed a typology of the self that was composed of both subjective and objective selves (Mead, 1934). The subjective self is without awareness of itself; it processes experience. The subjective self has also been referred to the "I" in the English language. The subjective self is most evident in our habitual routines and in our intentional behavior (Bruno, 1992). The objective self includes self-awareness. The objective self has also been referred to as the categorical self, the self concept, and the social self; it is the "me" in English language. In a sense, we possess a subjective self from an early age: Infants act on the world, have emotional states, and respond to social and physical stimuli. It is not until we develop the capacity to reflect on ourselves that we begin to develop objective self-awareness (Meyer & Salovey, 1997, p.41).

The sociologists Cooley and Mead both emphasized that our relationships with others are pivotal to how we come to describe ourselves – and thus develop a self-concept (Mead, 1934). Social psychologists have determined that those individuals who rely excessively on others' evaluations of themselves rather than on their own self-reflections are more prone to experience low self-esteem and their sense of well-being is fragile. As a result, they also experience more frequently mood swings and negative emotions (Higgins, Loeb, and Moretti, 1995). However, it is also important to realize that the self-concept often exists independent of the self as perceived by others. For example, in an extreme case, an individual may have a negative self-concept and be perceived as able and attractive by most other people (Bruno, 1992).

2.1.2.3 Neisser's model of tripartite self

Neisser's (1992) tripartite self model includes the following three aspects: the ecological self, the extended self, and the evaluative self. The ecological self is much like a subjective self in that the emphasis is on the individual engaged in transactions with those features of the environment that permit or afford interaction. Thus, the individual interacts with an environment that is bidirectional: What happens is a joint function of what we can do with the environment and what the environment provides us as an accessible structure (Gibson, 1982). Neisser's concept of a self extended in time is intended to address the fact that we are not only in a relationship to a present environment but are also concerned with the past and the future. The extended self permits us to imagine the possible, not just the actual. The preschooler's imaginary play is an early example of an extended self. The extended self also facilitates the development of schemas and scripts: We have a readiness for how to respond to a new environment based on some of its similarities to what we have learned as interaction strategies in former environment (Meyer & Salovey, 1997, p.43). According to Neisser, the evaluative self is a goal-directed self, and as a consequence, our environmental interactions are motivated toward some outcomes and away from others.

Neisser's model of the tripartite self helps us to look at emotional experience within the individual as it unfolds (a) in a physical and social environment (the ecological self), (b) relative to a temporal framework (the extended self), and (c) in response to the standards and values of the family and societal context (the evaluative self). In sum, the concepts of the ecological, extended, and evaluative self permit us to look at functional interactions between individuals and their social and physical environments.

This triple concept of self also gives us a conceptual tool to look at how individual differences may manifest themselves in why Person A feels self-efficacious in a particular social situation. The point to be made here is that the ostensibly similar social situation is not experienced *transactionally* as uniform. The social situation becomes a dynamic experience and functionally varies according to how Person B's multifaceted self engages in it. Emotional competence then becomes linked to how a particular multifaceted self experiences self-efficacy in particular transactions. This implies some degree of inconsistency or variability in emotional competence (Baumeister, 1993).

2.1.2.4 Self-concept and identity diffusion

Erikson (1959) says, the adolescent integrates and makes more internally consistent the various identities he has accumulated and is in the process of accumulating. An adolescent, for example, typically identifies with or imitates a number of different key individuals – his parents, a sibling, a close friend, etc.. "Identity diffusion" usually occurs when a young person tries to identify with multiple roles. The sorting, selecting of roles sometimes burdens a young person since many of the roles and identities are inconsistent and contradictory. Kenneth Kenniston (1965) pointed out that identity problems also result from the young person's attempt to span the gulf between childhood and adulthood while immersed in an intervening social environment that is disconnected from both of these periods and is also characterized by internal instability.

Unresolved problems of identity diffusion can lead to an identity crisis, which Peter Madison (1969) describes as a developmental crisis brought on by an individual's realizing that his personal qualities and capabilities are incompatible with the social roles available to him in his present or anticipated situation. Many identity crises occur in relationship to career choice, as when a student finds that he simply cannot reach social role requirements by all means, for example, he cannot conquer higher mathematics to realize his dream of being a math scientist. When the identity crisis comes on unexpectedly and very suddenly, the individual suffers an identity shock, in which he is confronted by a fundamental challenge to his self and social roles, which had been successfully integrated up until that moment. Still another type of identity crisis noted by Madison is a state he terms nonbeing, in which the individual's sense of identity is severely threatened and he fears that the self he knows will not continue to exist (Watson & Lindgren, 1979, p.551).

Problems faced by young girls and women: Fear of success The trait that has been identified with avoidance of competition and other success-oriented activity is what Matina S. Horner (1972) has termed fear of success. Horner maintains that the possibility of success, especially in competition with males, threatens women's self-esteem and feelings of feminity, and hence arouses anxiety lest one be rejected by others. Theoretically, a young woman has as much freedom as a man to select an occupational goal and prepare for it, but the social and psychological realities of life posed problems that are not easily resolved. Indeed, many female adolescents and adults avoid competing with males on issues that affect later success in a career.

2.1.2.5 Impact on achievement

Analyses of research studies in psychiatry and psychology indicate essentially that the student's ability to utilize the power to learn is determined by their self-concept or their perception of their world including personal goals, purposes, and values (Jones, 1992b). Teachers have long noted that children with a positive self-concept achieve better and perform the role of student better than a child with a negative self-concept. Students with negative self-concepts often become underachievers in school (Jones, 1992a).

In examining the self-concept factors of academic ability and academic achievement, it is necessary to consider the sense of personal control the child feels over his/her learning. A considerable body of research indicates that those children who possess awareness that they must be responsible for their own learning by taking an active role in that process are the most successful students (Reid, 1988). The child's sense of personal control is developmental and highly related to cognitive advances, and is often determined by assessing his/her abilities and perspectives in metacognition, motivation, and attribution (Jones, 1992a).

Metacognition   Metacognition refers to both the knowledge about cognition and the regulation of cognition (Reid, 1988). A number of self-control factors found in successful students include possession of metacognitive skills or the ability to self-plan, self-organize, and self-monitor during the learning process.

Motivation   "The relationship between self-concept, motivation, and achievement has long been an integral part of humanistic and open education program" (Biehler & Snowman, 1986, 519). "Motivation is the inner force that moves a person to take action toward a specific end." (Levine, 1989, 210) Motivation and metacognition interact to assist the student in evaluating their own behavior in numerous situations through perspective-taking and social comparisons, and in providing patterns of behavior to allow them to function in socially appropriate ways (Jones, 1992a).

Attribution   In school settings attributions are children's explanations and/or inferences about the causes of their academic, behavioral, or social performances and their evaluations. Attributions are formed over time and result from experiences within learning contexts and represent students' ideas about their control over their learning as well as achievement (Reid, 1988). Successful students usually attribute success to their effort and ability, while underachieving students attribute success to luck. Successful students attribute their failures to lack of effort and renew their efforts to master the task, while unsuccessful students generally attribute their failure to bad luck or task difficulty (Jones, 1992a). The causal attributions that individuals make about the success or failure of their actions (Weiner, 1994) influence motivation and performance largely through the mediational role of self-efficacy (Bandura, 1995; Schunk, 1991, 1994). In general, students who are confident of their ability intensify their efforts when failure occurs and persist until they succeed (Pajares, 1996; Pajares & Kranzler, 1995).

2.1.3 Hypothesis on Math self-concept

It was the researcher's interest to find out whether the relationship between environmental factors (positive and negative factors) and academic emotions (positive and negative emotions) was mediated by math self-concept. Therefore, the hypothesis on math self-concept was projected as the following:

- *Hypothesis 1:* *Math self-concept mediates the relationship between environmental factors and academic emotions.*

The hypotheses on the interrelations between math self-concept and achievement emotions, goals, and learning strategies, etc. were formulated after the literature review of the relevant theories in the following chapters.

2.2 Achievement motivation

2.2.1 Definitions of "achievement", "achievement motivation", and "need for achievement"

The study of motivation centers on the question of why people initiate, terminate, and persist in specific actions in particular circumstances (e.g., Atkinson, 1958; Mook, 1986). The guiding nominal definition of achievement (modified from Smith, 1969) is: achievement is task-oriented behavior that allows the individual's performance to be evaluated according to some internally or externally imposed criterion that involves the individual in competing with others, or that otherwise involves some standard of excellence (Spence, & Helmreich, 1983, p. 12). Achievement-related motivation refers to the personality factors that come into play when a person undertakes a task at which he will be evaluated, enters into competition with other persons, or otherwise strives to attain some standard of excellence (Smith, 1969, p.1). Need for achievement (n Ach) refers to motive to work persistently, energetically, and eagerly at the accomplishment of tasks and to seek new tasks to accomplish (Watson & Lindgren, 1979, p.573).

Under need for achievement a variety of motivational dispositions and cognitive assessments of the situation are activated and influence a person's tendency to behave (e.g., his tendency to work more or less hard, to persist or to give up and turn to another activity, his thoughts of doing well or poorly, and his physical manifestations of stress). Specifically, the determinants of performance dealt with include approach and avoidance motives (the motive to achieve, and anxiety about failure or the motive to avoid failure), expectations (subjective estimates of the likelihood that one's efforts will lead to success or failure), and the incentive values of success and failure (e.g., the pleasure of winning, or making a good grade, or executing a difficult task flawlessly, or not being good at something, of doing less well than one's companions) (Smith, 1969, pp.1-2).

2.2.2 Development of achievement motivation

2.2.2.1 Three stage development of achievement motivation

Veroff (1969) distinguishes between autonomous achievement motivation which involves internalized personal standards of excellence and social achievement motivation which involves external standards of excellence. A particular individual normally develops both kinds of achievement motivation, however, some persons are more internally oriented, focusing much on internal standards; whereas other individuals are more externally oriented, responding to external standards actively. Successful completion of the preceding stage is the prerequisite for satisfactory development in the next stage. Motivational typologies vary as a function of different stages of completion

- *Stage 1: Autonomous achievement motivation - early childhood*

Veroff (1969) describes that an early period (approximately the first 6 years) is critical to the development of autonomous achievement motivation. In this period exploration and coping lead to mastery and a sense of effectiveness. If the development of autonomous achievement motivation during the early years is prevented, or if emphasis on social comparison is introduced prematurely, then a mature form of achievement motivation can never develop.

- *Stage 2: Social achievement motivation – elementary school years*

During the elementary school years, the child learns to respond to external standards of accomplishment. The child is helped to define both the difficulty of tasks and his own level of ability by comparing himself with other children, and by becoming aware of external standards. A disposition to strive for achievement in social settings is acquired through a certain amount of success in competition with these social standards.

- *Stage 3: Integration of autonomous and social achievement motivations – adolescence*

In a third stage of development, the adolescent learns how to integrate the two kinds of achievement orientations. He must have the capacity to rely on internal standards and to respond to external standards in different circumstances appropriately.

2.2.2.2 Childrearing practices and strength of motives

Motives are viewed by Smith as learned dispositions to strive to attain positive incentives and to avoid negative incentives. Individual differences in the strength of motives are attributed to differences in childrearing and other learning experiences in the early life of the child. Smith's analysis of the childrearing values indicates that parental values for independence and achievement do not necessarily go together. That is, a parent can value independence highly and not achievement, or vice versa, which raises the possibility that these two aspects of childrearing may make different contributions to the development of achievement motivation. Parental values for achievement are not unitary. The achievement value items fall into two distinct groups labeled "assertive achievement" and "conscientious achievement," which are similar to competitive achievement and individual achievement. (Smith, 1969)

Rosen and D'sAndrade (1975) found in their study of the psycho-social origins of achievement motivation that achievement training contributes more to the development of achievement motivation than does independence training. The general pattern that emerges from this and other studies of n Ach in childhood is that boys and girls are more inclined to engage in achievement-oriented behavior if their parents encourage it by expressing high expectations and at the same time permitting the freedom and independence which enable children to try out their skills and develop problem-solving strategies (Watson & Lindgren, 1979).

2.2.3 Development of achievement motivation theories

Individual differences in intrinsic achievement motives have been the subject of extensive theorizing and empirical investigation by psychologists and other behavioral scientists. Early achievement theorists recommended by Spence and Helmreich (1983) in their book "Achievement and Achievement Motives: Psychological and Sociological Approaches" include Murray, Atkinson, McClelland, and their colleagues (Arkinson, 1958; Atkinson and Raynor, 1974; McClelland, Atkinson, Clark, and Lowell, 1953, Murray, 1938). Eccles' expectancy-value model (1983) shifted attention away from motivational factors to cognitive factors in determing achievement behaviors, which was a new perspective to look into achievement motivation. Spence and Helmreich (1983) are also listed below as representatives of the 80s due to their contribution to achievement motivation theory. In the last twenty years, achievement motivation theory turned to achievement goal analysis. The representatives are Dweck (1986), Elliot and Church (1997), and Elliot and McGregor (2001), etc.

2.2.3.1 Achievement motivation according to Murray

Murray conceived of personality as a series of needs, described as an "organic potentiality or readiness to respond in a certain way under given conditions" (1938, p.60). Among these needs is the need to achieve, which Murray described as "the desire or tendency to do things as rapidly and/or as well as possible... to accomplish something difficult. To master, manipulate and organize physical objects, human beings, or ideas.... To overcome obstacles and attain a high standard. To excel one's self. To rival and surpass others" (1938, p.164).

Influenced by psychoanalytic thought, Murray postulated that needs are largely unconscious; accordingly, he devised a projective instrument, Thematic Apperception Test (TAT), to assess them. The TAT consists of a series of ambiguous pictures of one or more people about whom test respondents are asked to tell a story. The fantasy material is then coded for the presence of imagery relating to various needs. Murray also used the TAT to assess need for achievement.

Spence & Helmreich (1983) have been especially critical of the TAT's low reliability, noting that respondents' stories are overly responsive both to the particular pictorial material used to elicit them and to the situational conditions under which the test is taken, and that scores derived from respondents' stories are not stable from one testing occasion to another (p.38). Spence and Helmreich (1983) are even more critical about the scoring method, which presupposes that achievement motivation is unifactorial, namely a single, broad disposition influencing a variety of behaviors in achievement-related situations (p.38). Several factor-analytic studies (e.g., Jackson, Ahmed, and Heapy, 1976; Veroff, McClelland, and Ruhland, 1975) have revealed the presence of a number of more or less independent factors.

2.2.3.2 Expectancy-value theory of Atkinson

McClelland and Atkinson were in line with Murray that motives are dispositional tendencies to strive for a certain kind of satisfaction. They further conceived of motives as having activating, affective, and goal-oriented properties. Atkinson describes both motives in general and the motive to achieve in particular: A motive is conceived as a disposition to strive for a certain kind of satisfaction, as a capacity for satisfaction in the attainment of a certain class of incentives. The names given motives – such as achievement – are really names of classes of incentives which produce essentially the same kind of experience of satisfaction (for example, in the case of achievement motive): pride in accomplishment…The general aim of one class of motives, usually referred to as appetites or approach tendencies, is to maximize satisfaction of some kind. The achievement motivation is considered a disposition to approach success (Atkinson, 1966, p.13).

- *Expectancy-value theory proposed by Atkinson (1957)*

Atkinson (1957) introduced his expectancy – value theory incorporating the stable personality characteristics of achievement motive. According to this theory, the strength of the achievement motive (the tendency to achieve) in any achievement-oriented context is determined by the sum of two tendencies with opposing signs: (1) The tendency to approach success, manifested by engaging in achievement-oriented activities; (2) The tendency to avoid failure, manifested by not engaging in these activities.

The strength of each of these opposing tendencies is determined by three components: (1) The motive to approach success or the motive to avoid failure; (2) The expectancy (probability) that an achievement-oriented act will result in success or the probability that it will result in failure; (3) The incentive value of success or the incentive value of failure.

The expectancy-value label came from the latter two variables of the three components. According to Atkinson (1957), the motive to approach success is an individual-difference variable. The motive to avoid failure, also called fear of failure, is proposed as a separate dispositional tendency that, like the motive to approach success, is a stable personality characteristic that has been acquired as a result of past experience. The second component determing the tendency to approach success or to avoid failure is expectancy, defined as the probability that engaging in an achievement-oriented activity will result in success or in failure. The third component, incentive value of success or failure, has been described by Atkinson as the degree of anticipated satisfaction or pride in succeeding at a task or the degree of anticipated shame in failing.

The three components associated with the tendency to approach success and with the tendency to avoid failure – motive, expectancy, and incentive – are assumed to combine multiplicatively to determine the strength of each of these tendencies. These two tendencies (given opposite signs), in turn, determine the strength of the resultant achievement motivation, or the tendency to achieve.

- *Elaborations of the theory:*

Extrinsic rewards (T_E) added to the original theory Atkinson and his colleagues revised and modified the original expectancy-value theory in order to improve the theory's predictive utility. According to the original expectancy-value theory, individuals would stop all achievement-related activities if their tendency to approach success is weaker than their tendency to avoid failure. However, in reality, many individuals are still busy with achievement-related activities even if they are more failure-avoidant oriented. To remedy this deficiency, Atkinson (1974) added another construct to the theory: the tendency to seek extrinsic rewards (T_E). This tendency combines additively with the tendencies to approach success (T_S) and avoid failure (T_{AF}), so that $T_A = T_S - T_{AF} + T_E$. Spence and Helmreich (1983) believe that intrinsic and extrinsic motives may be related in a complex manner and do not necessarily add together in any simple way to form a resultant motivational state. It does seem reasonable to assume, however, that extrinsic motives may buoy up achievement – oriented efforts in those whose intrinsic achievement motivation (or the tendency to approach success) is week and/or in those whose fear of failure is strong.

Motive to avoid success added to the original theory Another addition to expectancy – value theory is Horner's (1968) introduction of motive to avoid success. This motive (also commonly identified as fear of success) is described as a stable dispositional tendency, acquired relatively early in life, to become anxious about achieving success. Like fear of failure, fear of success is postulated to reduce resultant achievement motivation (or the tendency to achieve) and hence to inhibit achievement – related behavior.

Horner (1968) stated that women are particularly likely to develop fear of success since their sex – role socialization leading them to believe that achievement strivings are incompatible with femininity and that women incur a number of social penalties if they violate role expectations by attempting to become successful. Horner's hypotheses quickly commanded both popular and scientific attention and led to many empirical studies. Subsequent studies have produced conflicting outcomes (Condry and Dyer, 1976; Tresemer, 1977). Whereas some investigators have reported significantly more fear of success in women than in men, many others have reported either the reverse or no differences between the sexes.

2.2.3.3 Intrinsic and extrinsic motives of Spence and Helmreich

Ambition and the drive to achieve excellence are widely recognized as crucial ingredients in successful attainment. Wide individual differences in accomplishments may be observed even if people have similar education, ability, social background, and opportunity. Therefore, the intensity and the nature of their achievement-related motives must be taken into account to understand what makes some people more successful than others. They view achievement motivation as a cluster of interacting factors rather than as a single, unitary dimension (Spence, & Helmreich, 1983).

- *Intrinsic versus extrinsic motives and goals*

In the classical tradition of Murray and Atkinson, Spence and Helmreich regard the conception of intrinsic achievement motivation as the enjoyment of achievement-related activities and of striving toward performance excellence. Extrinsic achievement motivation is the desire for the tangible or intangible rewards that are often obtained as a consequence of successful performance. It is possible, of course, for a single set of behaviors to be driven simultaneously by both intrinsic and extrinsic motives (Spence, & Helmreich, 1983).

- *Intrinsically and/or extrinsically driven individuals*

Spence and Helmreich further distinguished individuals with strong extrinsic motives from individuals with both intrinsic and extrinsic motives. Extrinsically driven individuals usually do not have a supportive environment to meet their achievement needs and therefore their intrinsic motives are weak. Their effort invested in achievement-related activities can be shifted to a new direction if external incentives are re-directed. For persons with some degree of intrinsic motivation besides extrinsic motivation, extrinsic motivation has the possibility to bring out some extrinsic outcomes, whose function is positive in the sense that external feedback and reward provide information about the quality of their performance and personal competence and serving as ego-gratifying recognition of their accomplishments (1983).

In actuality, most people's achievement efforts are probably spurred by a number of interacting motives that vary in strength and saliency across individuals, and within individuals, across situations. Similarly, the consequences of successful performance may have multiple meanings and values to their recipients. Although these various aspects of motives and rewards may in practice be difficult to disentangle and measure separately, it is imperative that distinctions be maintained since they may have different implications for achievement behaviors (Spence, & Helmreich, 1983, p. 17).

Intrinsic motivation implies that performance is self-initiated, self-sustaining, and self-rewarding; whereas extrinsic motivation implies that performance is externally driven and is likely to be extinguished or diminished in the absence of reward (e.g., Kazdin and Bootzin, 1972). For these reasons, parents and teachers would be well advised to encourage the development in their charges of intrinsic motivation for performing desirable behaviors and to use extrinsic rewards judiciously (Spence, & Helmreich, 1983).

2.2.3.4 Expectancy-value model of Eccles

Expectancy-value theory of Atkinson (1957) should be differentiated from expectancy-value model of Eccles (1983). Atkinson's theory focuses on individual differences in the motive to achieve and on the effects of subjective expectancy on both this motive and the incentive value of success. Eccles' theory has shifted attention away from motivational constructs to cognitive constructs, such as causal attributions, subjective expectancies, self-concepts of abilities, perceptions of task difficulty, and subjective task value. In another word, Eccles' model, most directly influenced by theories in which the constructs of expectancy and value are prominent, focuses on the role of cognitive rather than motivational factors in determining achievement behaviors.

The Eccles' model itself is built on the assumption that it is not reality itself (i.e., past successes or failures) that most directly determines children's expectancies, values, and behavior, but rather the interpretation of that reality. The influence of reality on achievement outcomes and future goals is assumed to be mediated by causal attribution patterns for success and failure, the input of socializers, perceptions of one's own needs, values, and sex-role identity, as well as perceptions of the characteristics of the task. Each of these factors plays a role in determining the expectancy and value associated with a particular task. Expectancy and value, in turn, influence a whole range of achievement-related behaviors, e.g., choice of the activity, intensity of the effort expended, and actual performance (Eccles, 1983, p.81).

- *Expectancy-value model and its two components*

The first component is a *psychological component* in which the interactions of various cognitive factors at one point in time are specified; the second is a developmental component, which specifies the origins of individual differences in these psychological factors In the first component, the most immediate precursors of such performance variables as task choice and persistence are individuals' expectancies or subjective probabilities of success and the value they place on successful attainment. These expectancies and values, as they relate to children's school performance, are determined by such variables as the individuals' goals and self-concepts, their perceptions of parents' and teachers' expectations, their interpretations of the reasons (ability or lack of ability) for their past performance, and their perception of the difficulty of the task. (Eccles, 1983)

- *Psychological component one: expectancies*

Eccles et, al. (1983) proposed that expectancies are influenced most directly by self-concept of ability and by the students' estimate of task difficulty, indirectly by individual's interpretation of these past events, perceptions of the expectancies of others, and identification with the goals and values of existing cultural role structures.

Self-concept of ability Formed through a process of observing and interpreting one's own behaviors and the behaviors of others, self-concept of ability is defined as the assessment of one's own competency to perform specific tasks or to carry our role-appropriate behaviors. Research specific to math achievement has yielded a consistent and positive relation between perceptions of mathematical ability and plans to enroll in advanced mathematics courses (e.g., Armstrong and Kahl, 1978). Furthermore, females report lower estimates of their math abilities than do males. These differences do not emerge with any consistency prior to junior high school but are frequently found at and beyond junior high, despite the fact that, during elementary school and junior high school, females perform just as well as males in math (e.g., Fennema and Sherman, 1977; Heller, Futterman, Kaczala, Karabenick, and Parsons, 1978).

Perception of task difficulty Self-concept of ability appears to be the more critical construct. Perceptions of task difficulty, however, may influence self-concept of ability over time, students who see a subject or task as more difficult develop lower estimates of their own abilities for that subject or task (Spence & Helmreich, 1983, p.85).

Perception of others' expectations The achievement literature has documented the importance of parents' and teachers' expectations and attitudes in shaping students' self-concepts and general expectancies of success (Brookover and Erickson, 1975; Kaminski, Erickson, Ross, and Bradfield, 1976; Parsons, Frieze, and Ruble, 1976). However, the causal direction of this relation is unclear. When sex differences are evident, female students perceive their parents as having lower estimates of the females' abilities than do male students (Fennema and Sherman, 1977; Kaminski et al., 1976).

Causal attributions Attribution theorists have suggested another set of variables as important mediators of individual differences in expectancies and perceptions of both one's ability and the difficulty of the task (Frieze, et al. 1978; Weiner, 1974). According to these theorists, it is not success or failure per se, but the causal attributions made for either of these outcomes that influence future expectancies. Findings of attribution differences between male and females is not consistent (Dweck and Goetz, 1978; Parsons, Ruble, Hodges, and Small, 1976). Thus, whether sex differences in attributions mediate sex differences in achievement behaviors remains an open question.

Locus of control Locus of control is closely related to attribution theory. Dweck (1975) introduced the concept of academic learned helplessness to describe students who assume that they cannot control their failures. Eccles (1983) integrated locus of control into the expectancy-value model and gave such conclusions on locus of control: Empirical evidence has demonstrated the important mediating role of locus of control and learned helplessness for achievement-related behaviors; Sex differences have not been found consistently on either locus of control or learned helplessness; The mediating role of learned helplessness in accounting for sex differences in achievement has yet to established (Parsons, 1981).

- *Psychological component two: task value*

Eccles (1983) criticized Atkinson's (1964) definition of the value concept as too narrow, which is only based on objective task characteristics. She suggests that the overall value of any specific task is a function of three major components: (1) the attainment value of the task, (2) the intrinsic value of the task, and (3) the utility value of the task for future goals.

Eccles (1983) proposes that the value of a particular task to a particular person is a function of both the perceived qualities of the task and the individual's needs, goals, and self-perceptions. According to her three clusters of variables are important mediators: (1) sex roles, (2) perceptions of the cost of success, and (3) previous affective experiences with similar tasks.

Self-role identity Central to this line of argument is the assumption that sex-role identity and the sex stereotyping of particular achievement activities interact in influencing task value. Eccles suggests that the sex typing of the task will affect its perceived value only to the extent that one's sex role identity is a *critical* and *salient* component of one's self-concept. Eccles further argues that it is the sex differences in career goals rather than the sex typing of math courses per se that is the major mediator of the sex difference in the perceived value of advanced math courses (1983).

Cost of success or failure Variables influencing the cost of an activity include: the amount of effort needed to succeed, the loss of time that could be used to engage in other valued activities, and the psychological meaning of failure (Eccles, 1983).

Affective experiences Eccles stated that variations in affective experiences can take two quite different forms: (1) variations caused by overt, objective events like success, failure, and the responses or behaviors of major socializers such as parents and teachers, and (2) variations created by psychological factors such as causal attributions and individual differences in confidence or anxiety (1983).

- *Model testing*

The investigator (Eccles, 1983) tested the interrelationships among the cognitive factors specified in the model (***Figure 2.2.1***) and the contribution of these psychological variables to a measure with implications for actual achievement behavior: the students' intention to take additional math courses. These analyses confirmed the importance of children's self-concepts of ability, attributions for past performance, and perceptions of the beliefs of parents and teachers as determinants of expectancies, values, and course plans. Relatively few sex differences were found, but those that did appear confirmed the results of past investigations: females, in comparison with males, had lower confidence in their ability and perceived math as more difficult and less valuable. Such sex differences appeared to be related to parents' beliefs in the difficulty of math for their child.

2.2.3.5 Achievement goal framework

- *Two fundamental dimensions of achievement goals*

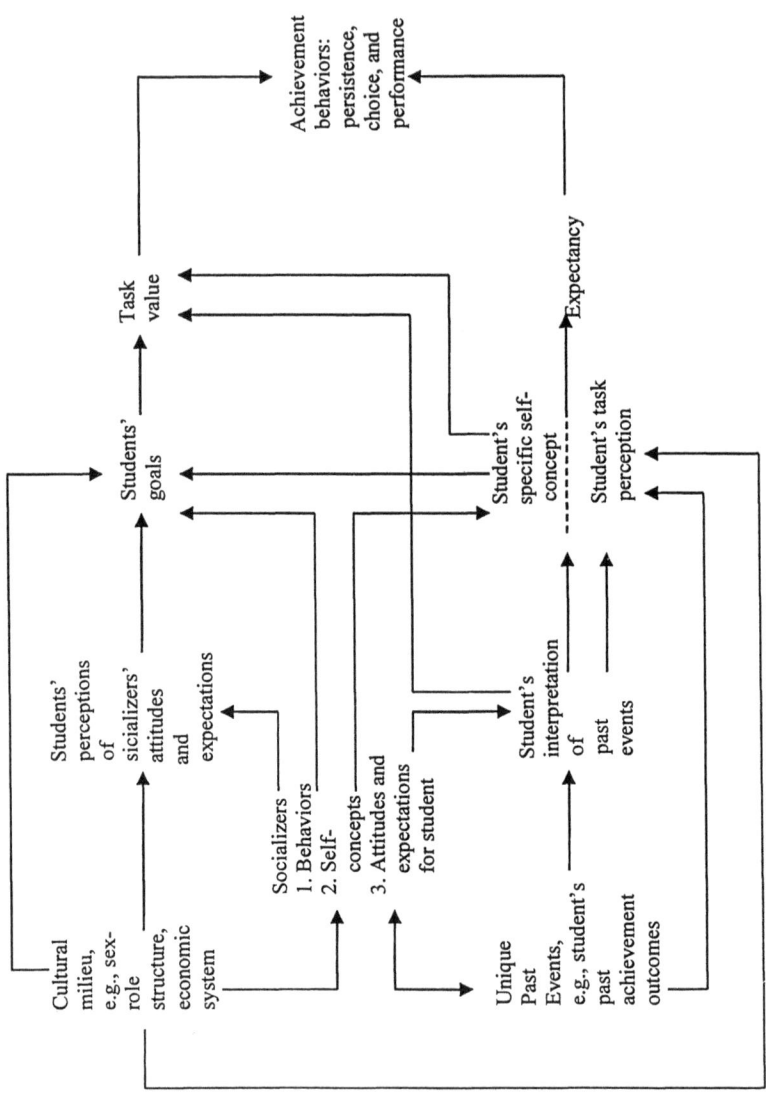

Figure 2.2.1: General Expectancy-Value and Developmental Model of Achievement Behavior

Source: Eccles, 1983, p. 80.

Competence according to Elliot and McGregor (2001) is at the conceptual core of the achievement goal construct. Competence and therefore, achievement goals, may be differentiated on two fundamental dimensions – according to how it is defined and according to how it is valenced. Competence is defined in terms of the reference or standard that is used in performance evaluation. Three different standards may be identified: absolute (the requirement of the task itself), intrapersonal (one's own past attainment or maximum potential attainment), and normative (the performance of others). That is, competence may be defined according to whether one has acquired understanding or mastered a task (an absolute standard), improved one's performance or fully developed one's knowledge or skills (an intrapersonal standard), or performed better than others (a normative standard). Absolute and intrapersonal competences share many conceptual and empirical similarities and often seem indistinguishable. The distinction between absolute/intrapersonal and normative standards was implicitly acknowledged in the classic conceptualization of achievement motivation, in that need for achievement was construed as a multidimensional construct that include doing well relative to task requirements and relative to others (McClelland, Atkinson, Clark, & Lowell, 1953; Murray, 1938).

The other fundamental dimension of competence is valence. Competence is valenced in that it is either constructed in terms of a positive, desirable possibility (i.e., success) or negative, undesirable possibility (i.e., failure) (Elliot and McGregor, 2001). Accumulating evidence indicates that persons process most, if not all, encountered stimuli in terms of valence and do so immediately and without intention or awareness (Bargh, 1997; Zajonc, 1998). Furthermore, this automatic, valence-based processing is presumed to instantaneously evoke approach and avoidance behavioral predispositions (Cacioppo, Priester, & Berntson, 1993; Forster, Higgins, & Idson, 1998). These approach and avoidance tendencies are present in infancy, appear to be grounded in the neuroanatomical structure of the brain, and likely represent part of the evolutionary heritage that humans share with organisms across the phylogenetic spectrum (Elliot & Covington, in press). The distinction between approach and avoidance forms of competence motivation was a central aspect of the classic conceptualization of achievement motivation (Atkinson, 1957; Murray, 1938).

According to Elliot and McGregor (2001), both dimensions – definition and valence – are integral to the competence construct and, therefore, must be viewed as necessary components of any and all competence-based forms of regulation, including achievement goals.

- *Dichotomous model of achievement goal framework*

Over the past 2 decades, a majority of the theoretical and empirical work conducted in the achievement motivation literature has used an achievement goal perspective (Elliot & McGregor, 2001). Achievement goals are viewed as purpose (Maehr, 1989) or cognitive-dynamic focus (Elliot, 1997) of competence-relevant behavior.

The primary emphasis has been on two goal types: mastery goals and performance goals (Dweck, 1986; Nicholls, 1984). Mastery goals are focused on the development of competence through task mastery, whereas performance goals are focused on the demonstration of competence relative to others (Elliot & McGregor, 2001). The valence dimension of achievement goals had not been included by that time. Each goal is presumed to provide a distinct perceptual-cognitive framework in achievement settings, and the two goals have been shown to lead to a differential pattern of processes and outcomes (see Ames, 1992; Dweck, 1999; Urdan, 1997). Mastery goals are more related to Veroff's (1969) *autonomous achievement motivation* which involves internalized personal standards of excellence; Performance goals are more related to *social achievement motivation* which involves external standards of excellence.

- *Trichotomous model of achievement goal framework*

Recently, Elliot and his colleagues (Elliot & Church, 1997; Elliot & Harackiewicz, 1996) proposed that the original mastery-performance goal dichotomy be revised to include the distinction between approach and avoidance motivation. Specifically, they bifurcated the performance goal construct to form performance-approach and performance-avoidance goals. Empirical research on this trichotomous framework has yielded strong support; factor analysis work has validated the independence of the three goal constructs (Elliot & Church, 1997; Middleton & Midgley, 1997; Skaalvik, 1997; Vandewalle, 1997) and the goals have been linked to differential pattern of antecedents and consequences (Elliot, 1999). Performance-approach goals appear to have no effects or even positive effects on children (e.g., Elliot & McGregor, 1999; Middleton & Midgley, 1997; Skaalvik, 1997), and performance-avoidance goals appear to have negative effects on children (e.g. Elliot & Harackiewicz, 1996; Grant & Dweck, 2003; Middleton & Midgley, 1997).

The Chinese goal orientation questionnaire study based on the three goal constructs also validated the independence of the three goal constructs (Lau & Lee, 2008). Lau and Lee (2008) also found that mastery goals were the most adaptive for students' learning. It was positively related to students' perceived mastery-oriented classroom environment and self-efficacy. Performance approach goals were also found to have a positive relationship with students' perceived classroom environment and self-efficacy, though the correlation coefficients were not as high as mastery goals. The relationship between performance-avoidance goals and other variables in the study were different from the hypothesis of the goal orientation theory. Contrarary to being a maladaptive type of goals in the goal orientation theory, performance goals were positively and significantly correlated with mastery goals, perceived classroom environment, and self-efficacy. Shih (2005) also found that mastery goals and performance-avoidance goals were positively related among Taiwan students and argued that the significant correlation between performance-avoidance goals and mastery goals was due to the shared variance between the two components of performance goals.

- *Two times two achievement goal framework*

Elliot and McGregor further bifurcated mastery goals to create mastery-approach and mastery-avoidance goals. Factor analysis validated the independence of the four goal constructs:

mastery-approach, mastery-avoidance, performance-approach, and performance-avoidance goals (Elliot and McGregor, 2001). Distinct empirical profiles for each of the achievement goals were revealed; the pattern for mastery-avoidance goals was more negative than that for mastery-approach goals and more positive than that for performance-avoidance goals.

- *The connection between trichotomous model and two times two factor model of achievement motivation*

The trichotomous achievement goal framework (Elliot & Church, 1997; Elliot & Harackiewicz, 1996) comprises mastery-approach goals (in which competence is defined in absolute/intrapersonal terms and is positive), performance-approach goals (in which competence is defined in normative terms and is positively valenced), and performance-avoidance goals (in which competence is defined in normative terms and is negatively valenced). The remaining cell of 2 times 2 frameworks is mastery-avoidance goals (in which competence is defined in absolute/intrapersonal terms and is negatively valenced).

		Definition	
		Absolute/ Intrapersonal (Mastery)	Normative (Performance)
Valence	Positive (Approaching success)	Mastery approach goal	Performance approach goal
	Negative (Avoiding failure)	Mastery avoidance goal	Performance avoidance goal

Figure 2.2.2: The 2 Times 2 Achievement Goal Framework

Source: Elliot &McGregor, 2001

Note: Definition and valence represent the two dimensions of competence. Absolute/intrapersonal and normative represent the two ways that competence can be defined; positive and negative represent the two ways that competence can be valenced (Elliot & McGregor, 2001).

- *The extension of 2 times 2 factor achievement motivation based on the trichotomous model: mastery-avoidance goal*

The concept of a mastery-avoidance goal may seem counterintuitive (Elliot & McGregor, 2001). In the mastery-avoidance goal construct, competence is defined in terms of the absolute requirements of the task or one's own pattern of attainment and incompetence is the focal point of regulatory attention. Prototypic examples include perfectionists who strive to avoid making any mistakes or doing anything wrong or incorrectly (Flett, Hewitt, Blankstein, & Gray, 1998; Pintrich, 2000b). The focus is on avoiding a negative possibility.

Conceptually, mastery-avoidance goals differ from mastery-approach goals in terms of the valence of competence, from performance-avoidance goals in terms of the definition of competence, and from performance-approach goals in terms of both the definition and valence of competence (Elliot & McGregor, 2001).

Empirical predictions regarding the antecedents and consequences of mastery-avoidance goals are difficult to generate for two reasons (Elliot & McGregor, 2001). First, the mastery component of the goal is likely to emerge from optimal antecedents and to facilitate positive consequences (like mastery-approach goals), but the avoidance component is likely to emerge from nonoptimal antecedents and to have negative consequences (like performance-avoidance goals). Second, it is impossible to know the relative strength of the two components when combined or the precise way in which each component functions in conjunction with the other.

2.2.4. Some issues on achievement motivation

- *Gender difference in achievement expectancy and achievement motivation*

Crandall (1969) reports that sex differences in expectancy show up in the elementary school years and are still present in college-age men and women. It is unclear that why girls should have lower expectancies of intellectual and academic accomplishment than boys when their performance is as good as, or even better than those of boys. Smith (1969) gave the following explanations for girls' low achievement expectancy. (1) Girls want to fulfill socially desirable estimates for females: being modest and diffident; (2) Girls have different values for intellectual and academic attainment; (3) The achievement expectancies of boys and girls are affected in different ways by positive and negative reinforcement of their efforts at a task.

Horner (1968) introduced the concept of *motive to avoid success*, which was added to expectance-value theory. This motive (also commonly identified as fear of success) is described as a stable dispositional tendency, acquired relatively early in life, to become anxious about achieving success. Like fear of failure, fear of success is postulated to reduce resultant achievement motivation (or the tendency to achieve) and hence to inhibit achievement – related behavior. Horner stated that women are particularly likely to develop fear of success since their sex – role socialization leading them to believe that achievement strivings are incompatible with femininity and that women incur a number of social penalties if they violate role expectations by attempting to become successful.

Concerning achievement goals, some studies have not reported a significant sex difference in the endorsement of mastery goals (e.g., Patrick, Ryan, & Pintrich, 1999; Ryan & Pintrich, 1997), but studies that have found a difference consistently indicate that girls focus more on learning and mastery than do boys (e.g., Ablard & Lipschutz, 1998; Meece & Holt, 1993; Nolen, 1988). Research has also shown that boys are more performance oriented than girls (e.g., E. M. Anderman & Midgley, 1997; Roeser, Midgley, & Urdan, 1996; Ryan, Hicks, & Midgley, 1997; Stipek & Gralinski, 1996). When children are concerned with mastery rather than with performing better than others, they are more likely to apply effective learning strategies (Elliott & Dweck, 1988; Grant & Dweck, 2003; Mueller & Dweck, 1998; Wolters,

Yu, & Pintrich, 1996). Whether achievement motivation being a gender issue or being influenced by other factors was discussed in details in Discussion part.
- *Conflicts between affiliation and achievement*
According to Watson & Lindgren (1979), affiliation versus achievement is an often found conflict among adolescents. It is the adolescent's need for affiliation – n Aff – that leads him to become involved in peer group. N Ach causes him to become involved in activities that provide him with feedback as to his competence. Achievement activities are inconsistent with n Aff, which focuses on such values as being accepted by and getting along with others, conforming to their expectations, etc... An adolescent who wants to take care of the two needs usually experiences conflict when personal interests and group demands compete for his time and attention. The group member who is preoccupied with self-improvement is likely to be regarded with some distaste by his peers.

2.2.5 Motivation and achievement in collectivist and individualist cultures

2.2.5.1 Definitions of the two cultural syndromes

- *Need to examine the two cultural syndromes*

Individualism and collectivism are cultural syndromes that differentiate the main cultures of Europe and North America, north of the Rio Grande from the main cultures of Africa, Asia, and the Pacific Islands. Individualists focus on the achievement of personal goals, by themselves, for the purpose of pleasure, autonomy, and self-realization. Collectivists focus on the achievement of group goals, by the group, for the purpose of group well-being, relationships, togetherness, the common good, and collective utility (Triandis, 1995).

Traditional (western) theories, suggest that the achievement motivation trait is developed through childrearing processes which emphasize independence, mastery and competitiveness (McClelland et al., 1953). Cross-cultural researchers in particular have reported that in many cultures achievement behavior does not confirm with the western prototype described by McClelland and his associates (cf. De Vos, 1968; Maehr, 1978), It has been suggested that achievement should be understood and defined in terms of the social and cultural context of individuals (Salli, 1995).

The findings (e.g. Gallimore, Boggs, & Jordan, 1974) suggest that different cultural groups hold varying beliefs and values about achievement, and achievement may mean different things to different people. Evidence from Salili's (1995) study suggests that some aspects of human achievement behavior may be universal (e.g. dimensions of achievement are the same for individualists and collectivists, namely affiliative success and individualistic success), while others vary under the influence of sociocultural and situational factors (e.g., There is a close association between succeeding in personal social life, family social life, and academic work, career and wealth for the Chinese. For the British, however, these areas of success are unrelated).

Achievement motivation among the Chinese is mediated more by affiliative concerns than personal needs and that their orientation towards achievement may be different from that of western nations (Salili, 1995). Salili (1995) further argues that Chinese students' achievement motivation tends to be affiliatively based. They also have a more adaptive learning style, attributing their success and failure to internal and controllable causes and they tend to be learning goal oriented.

Besides the reason to understand achievement in cultural context of individuals, it is necessary to examine carefully the two cultural syndromes in the current research project since: (1) the measures for the study were from individualist cultures (the United States and Germany); (2) the measures were administed to samples from the collectivist culture of China.

- *Central theme of cultural syndromes – Self: independent construal of self vs. interdependent construal of self*

A cultural syndrome consists of shared attitudes, beliefs, norms, roles, values, and other elements of subjective culture (Triandis, 1972), found among those who share a language, historical period, and geographic locations (Salili, 1995). It is organized around a theme. In individualist cultures the theme is the self as an autonomous and independent entity; in collectivist cultures the theme is the self as an aspect or part of one or more in-groups or collectives (Triandis, 1995). In the Chinese culture, self has two components that of *ta wo* (greater self) and *hsiao wo* (smaller self). The latter is referred to individual's own desires and actions for himself or herself (Hsu, 1985). Chinese emphasize greater-self which concerns with the well-being of others (e.g., family and society) (Salili, 1995).

According to Triandis (1995) social behavior among individualists is a function of attitudes, beliefs, personal habits, and self-definitions, while social behavior among collectivists is primarily a function of norms, roles, obligations, and group definitions. Geertz stated that the person with independent construal of self is viewed as "a bounded, unique, more or less integrated motivational and cognitive universe, a dynamic center of awareness, emotion, judgment, and action organized into a distinctive wholes and against a social and natural background" (1975, p. 48). The essential aspect of this view involves a conception of the self as an autonomous, independent person (Markus & Kitayama, 1991). The view of the interdependent construal of self and the relationship between the self and others features the person not as separate from the social context but as more connected and less differentiated from others. People are motivated to find a way to fit in with relevant others, to fulfill and create obligation, and in general to become part of various relationships (Markus & Kitayama, 1991).

The self-system has been shown to be instrumental in the regulation of intrapersonal processes such as self-relevant information processing, affect regulation, and motivation (see

Markus & Kitayama, 1991; Cantor & Kihlstrom, 1987; Greenwald & Pratkanis, 1984; Markus & Wurf, 1987, for reviews).
- *Fundamental syndromes: Complexity versus simplicity and cultural tightness versus looseness*

Individualism and collectivism can be conceptualized as reflections of two fundamental syndromes: cultural complexity versus simplicity and cultural tightness versus looseness. Cultural complexity is the most important dimension of cultural differences (Ember & Levinson, 1991), and contrasts the relatively simple social organizations of hunters and gatherers (e.g., the Eskimo) with the complex, modern, information cultures. Cultural tightness, as opposed to looseness, is characterized by severe sanctions when behavior does not follow ingroup norms. Tightness-looseness is situation specific. Individualism is maximal in cultures that are both complex and loose; collectivism is maximum in cultures that are simple and tight (Triandis, 1995). The Chinese culture is in the process of transformation, namely transforming itself toward a more complex culture and society.

2.2.5.2 Collectivism

- *Confucianism*

Central to Chinese cultural values is Confucian teaching and philosophy which dates back 2000 years to the Han Dynasty. Emperor Wu Ti created five colleges for official students to study five Confucian classics. As of this time potential officials were educated and selected based on mastery of Confucian ethics and Confucianism gradually became the official philosophy of the state (Fairbank & Reischauer, 1973). This philosophy taught people to exercise restraint over their desires and to distribute scarce resources equally among the group members. The most important assumption of Confucian philosophy is that "man is a relational being, socially situated and defined within an interactive context" (Bond, 1986, p. 215).

An individual's life is considered to be the continuation of his ancestors (Hsu, 1967), and his offspring's life the continuation of his own. One of the most important goals for an individual is "to do his best in continuing and making prosperous his family life" (Hwang, 1990, p.598). Chinese attach great importance to filial piety which emphasizes respect and unquestioning obedience and obligations towards one's parents. Thus, Chinese social behavior is characterized by a tendency towards collectivism (Hofstede, 1983). According to Hofstede, "Collectivism stands for a preference for tightly knit social framework in which individuals can expect their relatives, clan, or other members of this in-group to look after them in exchange for unquestioning loyalty. The fundamental issue addressed by this dimension is the degree of interdependence a society maintains among individuals" (1983, p. 83).

- *Confucianism and socialization for achievement*

Ho's (1986) review of literature on Chinese patterns of socialization; two areas of primary importance to Chinese parents are impulse control and academic achievement which are closely related to each other. Influenced by Confucian philosophy, the Chinese place great emphasis on education. Achievement through effort, hard work and endurance is highly emphasized and working hard and achieving beyond one's abilities is considered a virtue (Yang, 1986).

Salili and Ho (1992) examined the perceived attainment in a recent examination and satisfaction with performance. It was found that about 65 percent of students had low expectations of their own academic performance, while their mothers had generally high expectations of their children's achievement. Both mothers and students were dissatisfied with the examination results. Bottomley (1990) reported that 70 percent of the secondary and 90 percent of the college students in Hong Kong considered success in achieving their academic goals as the most of important source of their satisfaction. Stevenson and Lee (1990), in a large scale cross-cultural study, compared the context of achievement in Japan, Taiwan, and in the United States. They reported that Chinese families attach greater importance to their children's academic achievement than American families. Similar results were reported in comparative studies of United Kingdom and Hong Kong students. There is greater emphasis on good academic results in Hong Kong than in the United Kingdom (Winter, 1990).

2.2.5.3 Implications for emotion, motivation and achievement

- *Implications for emotion*

According to anthropologists Rosaldo (1984), Lutz (1988), and Solomon (1984), culture can play a central role in shaping emotional experience (Markus & Kitayama, 1991). Lutz (1988) claimed that "emotions can be viewed as cultural and interpersonal products of naming, justifying, and persuading by people in relationship to each other. Emotional meaning is then a social rather than an individual achievement – an emergent product of social life." (p. 5) Pride in one's own performance may be inhibited among those with interdependent selves (Markus & Kitayama, 1991). In line with assumption, Stipek, Weiner, and Li (1989) found that the Chinese were decidedly less likely to claim their own successful efforts as a source of pride than were Americans.

- *Differences in motivation: self-concept, self-esteem, self-efficacy, attributions*

Individualists have much more positive self-concepts than collectivists and collectivists have lower self-esteem than individualists. The low self-esteem of collectivists means that they will not want to be identified as standing out in relation to other members of the group (Triandis, 1995). Collectivists are lower on the attribute of self-efficacy, especially when they are low in ability (Oettingen, 1993). Earley (1993) carried out in China, Israel, and the United States a study of managers, and found that collectivists see the self as higher in self-efficacy when working with their ingroup, and individualists see the self as higher in self-efficacy when working alone.

Individualists attribute success and failure to ability much more than do collectivists, who focus more on effort. Mordkowitz & Ginsburg (1987) explain the academic success of Asian adolescents by their use of the "effort leads to success" attribution. Collectivists do not use the Performance = Ability times Effort conception as frequently as do individualists (Singh, 1981). They use a Performance = Ability + Effort conception. The focus on effort is desirable from the point of view of persistence, since effort is variable and under the control of the individual. Markus and Kitayama (1991) report studies that indicate that in Japan the tendency to de-emphasize the role of ability in explaining success can be found as early as the second grade.

- *Implications for motivation and achievement*

"Progress" as a similar form of achievement is different between individualist cultures and Confucian-influenced collectivist cultures. Individualists tend to see progress as consisting of radical improvements. Confucius taught a gradualist conception of progress, which is rather different from the Western view. The Confucian progress consists of using almost all that has worked in the past and making one minor improvement at a time, to see if that is a real improvement before making a second improvement and a third, and so on (Triandis, 1995).

Among those with interdependent selves, striving to excel or accomplish challenging tasks may not be in the service of achieving separateness and autonomy, as is usually assumed for those with independent selves, but instead in the service of more fully realizing one's connectedness or interdependence (Markus & Kitayama, 1991), fulfilling the expectations of the ingroup, for example, the family (e.g., Salili, 1995; Bond, 1986). In contrast, achievement for individualists is individual achievement, and is often seen as a means for self-glory, fame, and immortality. Individually oriented achievement motivation is viewed as a functionally autonomous desire in which the individual strives to achieve some internalized standards of excellence (e.g., Salili, 1995; Bond, 1986).

Collectivism has important implications for achievement. Family and group goals are given a higher priority and more importance than those of the individual (Hui, 1988). For collectivists, an achievement is group achievement, for the sake of the ingroup, or to show the superiority of the ingroup in relation to outgroups (Triandis, 1995). Similarly, academic success of the child brings a sense of pride to the entire family, while, academic failure is perceived as letting one's family down and causing them to lose face (Stigler, Smith, & Mao, 1985).

McClelland (1961) found that the level of various motives is a fairly direct reflection of the collectivist or group-oriented tradition of the Chinese. Yu (1974) found that achievement motivation was correlated with filial piety among collectivists. Salili and Ching (1992) in a study among high school students in Hong Kong who had just completed their General Certificate of Education examination, found both high and low achievers rated pleasing their parents as the most important reason for working hard. Yu and Yang (1987) in their studies among the Chinese in Taiwan found that Chinese achievement goals are for the benefit of the group, rather than the individual. Studies (Wilson & Pusey, 1982) provided evidence that the Chinese place greater importance on family and group goals than on individual goals in achievement and they are very concerned about loss or gain of their collective face in their pursuit of achievement. Markus and Kitayama (1991) reported that Japanese high school students' achievement motivation originated from a desire to meet the expectations of the group. "The groups is the family, and the child's mission is to enhance the social standing of the family by gaining admission to one of the top universities" (Markus & Kitayama, 1991, p. 241). Wilson & Pusey (1982) reported that Chinese people with strong collectivistic tendencies and face consciousness were the most achievement motivated.

However, they cautioned that with modernization the younger generations are adopting more and more western ideas and are moving towards individualism. Lau (1986) expressed similar concern and argued that the image of traditional Chinese culture may be overly idealized and simplified. The present picture of the Chinese culture is relatively unknown and may not be the same as the traditional image.

2.2.6 Hypotheses on achievement motivation

The current research focused on Chinese high school students' achievement goal orientation, namely whether they adopted mastery approach, mastery avoidance, performance approach, or performance avoidance goal; how their achievement goal related to math self-concept, achievement values, academic emotions, and environmental factors. However, the antecedents of their achievement motivation, namely the collectivistic cultural impact on achievement goal were not the focus of the study.

The hypotheses on achievement motivation under the whole framework of the study were projected as the following:

- ***Hypothesis 2.1:*** *Self-concept and self-efficacy positively correlate with approach goals and negatively correlate with avoidance goals.*
- ***Hypothesis 2.2:*** *Intrinsic value positively correlates with mastery goals; extrinsic value positively correlates with performance goals.*
- ***Hypothesis 2.3:*** *Positive environmental factors positively associate with approach goals; negative factors positively associate with avoidance goals.*

2.3 Affective development and academic emotions

2.3.1 Psychological perspectives on emotion

Psychological perspectives

According to LaFreniere (2000), psychological theories on emotion are categorized around three perspectives: (1) the psychodynamic tradition represented by Freud and Erikson; (2) cognitive and cultural approaches represented by Arnold, Lazarus, Averill, Oatley, and Weiner etc.; (3) developmental perspectives represented by Sroufe, Tomkins, Izard, Lewis, Michalson, Campos, Barrett, etc... In each of the three specific perspectives there are parallel or complementary theories. Therefore the representatives of a certain perspective are essentially representatives of their own specific theories under a general perspective.

The place of emotion in the study of human development can be understood from the "organizational" perspective. These two principles synthesized from the work of major developmentalists such as James Mark Baldwin (1897), Glick, (1992), and Piaget (1952), are unity and emerging complexity (Sroufe, 1996). By *unity* one means that the organism develops as a whole (Fogel, 1993; Gottlieb, 1991; Magnusson, 1988.). In the course of human development, for example, cognitive, social, and emotional developments are all part of the same process (Sroufe, 1996, p.36). *Emerging complexity* implies both that new, more complex behavior (and a new structure) emerges from what was previously present and that the new creations will show "emergent properties" (Fogel, 1993; Gottlieb, 1991; Waldrop, 1992). Development is characterized by *directionality* (toward increased complexity) and by *qualitative* change. Emerging complexity is inherent in development itself (Fogel & Thelen, 1987).

The current focus of the study is the dynamic processes between cognition and affect, namely the interaction between cognition and specific achievement emotions in the achievement setting of learning mathematics. Therefore the perspective taken in the current study is social-cognitive in nature. However, developmental perspective is applied in introducing the important concepts, e.g., emotional development, etc..

Interdependence of emotions and cognitions

Knowing is inherent in feeling (Kellerman, 1983). While the view of cognition leading emotions is in a sense valid, it is misleading in an important way, because the influence of emotion and emotional development on cognition would seem to be equally profound. Even granting that a cognitive process necessarily occurs in triggering an emotional reaction, this "cognition" is tempered by "affective memories" (Lazarus, 1991) – that is, past emotional experiences with that situation. Moreover, as Piaget and other developmentalists have suggested, affect is what cognition serves, and affective experiences alter cognitive structures. As tentative and infirm cognitive advances make possible emotional reactions, these experiences feed back to the cognitive system. As Piaget and Inhelder (1969) put it:

There is no behavior pattern, however intellectual, which does not involve affective factors as motives; but, reciprocally, there can be no affective states without the intervention of perceptions or comprehensions which constitute their cognitive structure. Behavior is therefore of a piece, even if the structures do not explain its energetics and if, vice versa, its energetics do not account for its structures. The two aspects, affective and cognitive, are at the same time inseparable and irreducible (p.158).

One cannot say that affect causes or precedes cognition, nor can one say that knowledge precedes affectivity (Piaget, 1962). Cognitive and socioemotional aspects of development are inseparable. Neither the cognitive nor the affective system can be considered dominant or more basic than the other (Sroufe, 1996, p.100).

The attachment theory provides an excellent illustration of the interplay between affect and cognition in development. In this position, attachment is an emotional construct; yet it is rooted in cognitive processes. While attachment itself is defined as an affective bond, there are "necessary conditions" for the emergence of this bond between infant and care giver (Sroufe, 1996). Ainsworth (1973) cites among these conditions a discrimination of the care giver from others and recognition of the person "as having a permanent an independent existence even when not present to perception" (p.28). Moreover, the evolution of the early attachment relationship to what Bowlby (1969/1982) calls a "goal-corrected partnership," in which the infant understands the caregiver's goals and can attempt to alter those goals, as well as his or her own behavior in terms of those goals, is obviously dependent on cognitive development. In fact, it is the apex of dyadic emotional regulation, a culmination of all development in the first year and a harbinger of the self-regulation that is to come (Sroufe, 1996, p.172).

2.3.2 Current situation of studying emotion

In recent years, there has been an increasing appreciation of the interrelatedness of emotion, motivation, and behavior, and of the role of emotion in normal development and social functioning. In the past emotions generally were seen as playing little causal role in normal social behavior. Emotions are now viewed as both a product of and a process in social interaction (Parke, 1994).

Despite the growing importance of the topic of emotion in developmental and personality psychology, research and theory pertaining to the role of emotion and its regulation in normal social functioning are limited (Bridges and Grolnick, 1995). For example, Hubbard and Coie noted the dearth of research on the relation of the ability to regulate emotions effectively to social competence with peers (1994). Much of the literature pertaining to emotion during childhood concerns children's understanding of others' emotions and when emotions are masked rather than the experience, expression, and regulation of emotions (Saarni, 1990). Of the limited work on the latter topics, most pertains to early socioemotional functioning (e.g., temperament and emotion in early relationships) (Bridges and Grolnick, 1995).

Eisenberg and Fabes (1992) argued that individual differences in emotional intensity and in regulatory capabilities play a pivotal role in a variety of aspects of social functioning. The way people react to and cope with emotions is likely influenced by cultural norms, namely how others in one's social world handle and respond to emotions. Moreover, the ways individuals experience, express, and cope with feelings would be expected to influence the quality of associated behavior, as well as long-term outcomes such as quality of social relationships and feelings of competence. Furthermore, it is probable that how individuals experience, express, and regulate their emotions influence the socialization process (Bridges and Grolnick, 1995).

2.3.3 Definitions, components of emotions, and emotional states

Definitions

Emotional development is sometimes also referred to as affective development. The term affect in psychology has become more or less synonymous with the word emotion. At other times *affect* is used to denote the expressive component (Oxford English Dictionary) or the subjective feeling component (Webster's International Dictionary) of emotion. Sroufe insisted that the use of the term emotion be restricted to the complex reactions, which includes cognitive, affective, physiological, and other behavioral components (Sroufe, 1996). A technical distinction is sometimes made indicating that affect is the display of an emotion in contrast to the emotion itself (Bruno, 1992).

Emotion is conceptualized by psychologists, Gerrig and Zimbardo (2002, p.394) as "a complex pattern of bodily and mental changes that includes physiological arousal, feelings, cognitive processes, and behavioral reactions made in response to a situation perceived as personally significant". Functionalist perspectives conceptualize emotions as responses that guide the individual's behavior and serve as information that helps the individual achieve goals (Bretherton, Fritz, Zahn-Waxler, & Ridgeway, 1986).

Components and functions of emotions

Components of emotions Emotions can be thought to have three components: cognitive-experiential, behavioral-expressive, and physiological-biochemical. The cognitive-experiential component comprises one's thoughts and awareness of emotional states (i.e., what most people refer to as "feelings"). The behavioral-expressive component comprises such domains as body movement, facial expression, posture, and gesture. The physiological-biochemical component comprises physical states and is reflected in such measures as brain activity, heart rate, skin response, and hormone levels. The physiological-biochemical component and the cognitive-experiential component are generally not visible to others (Dodge, 1989).

Greenberg and Snell (1997) defined emotion in neurobiological terms and claimed that emotion includes at least the following four components: (1) an expressive or motor component, (2) an experiential component, (3) a regulatory component, and (4) a recognition or processing component. Further, each of these components is recognized as involving particular neural/brain processes. The neurobiological defined components (Greenberg and Snell, 1997) includes a regulatory component, which is absent in Dodge's (1989) three components. The first component is the ability to express emotion through facial expression, body posture, and vocal tone and content. The second component is our conscious recognition of our emotions and feeling. These feeling states are normally expressed after language acquisition, by verbal report of experience. The third component of emotion is the regulation of emotion. The final component of emotion is the ability to recognize emotions in others by processing their facial expression, body posture (including head and eye movements), and vocal tone and speed (Greenberg and Snell, 1997).

Functions of emotions Functions of emotions are based on Darwinian Theory. That is, emotional reactions must be functional with respect to promoting the safety, environmental mastery, and, ultimately, reproductive success of the animal (Plutchik, 1983, p.223). According to Sroufe (1996), prominent functions of human emotions are (1) to communicate inner state to important others, (2) to promote exploratory competence in the environment, and (3) to promote adequate responses to emergency situations (p.16).

Differential emotions theory and emotional states

Positive and negative affect The characteristics of emotional responses may be determined from situations in which they occur, as well as from the general direction of the behavior that is displayed. Positive emotions are characterized by a tendency to *approach*. Enjoyment, satisfaction, and love all involve being attracted to objects or persons-wanting to retain them, remain with them, or stay in their proximity (Watson, & Lindgre, 1979, p.311). Negative emotions, are of two main types: fear and rage. Fear or anxiety response have an *away* from quality to them – that is, we want to remove ourselves from offending objects or persons or want them to depart from us. Rage responses have an *against* quality; when we are irritated or angry we want to destroy, humiliate, or injure the offending objects or persons. In very general terms, fear or anxiety responses are characteristic of us when we feel less than adequate to deal with the situation in question, whereas rage or anger responses are associated with a feeling of strength, power, and competence (Watson, & Lindgre, 1979, p.311).

Four emotional states Positive and negative affects are further developed to include four emotional states. According to Bruno (1992, pp 105-106), the four basic emotional states are derived from the two primary emotional dimensions. The emotional dimensions are bipolar in nature and are identified as (1) pleasure-distress and (2) excitement-calm. Pleasure-distress is called the *hedonic* dimension. Excitement-calm is called the *arousal* dimension.

The four basic states are, therefore, (1) pleasure with excitement, (2) distress with excitement, (3) pleasure with calm, and (4) distress with calm. The four basic states can be readily identified in infancy. They become more articulated and detailed in early childhood, and this is the process known as emotional development. By late childhood the individual is possessed of a complete emotional range. Emotional states such as ecstasy, joy, and delight reflect pleasure with excitement. Emotional states such as rage, anger, and anxiety reflect distress with excitement. Emotional states such as tranquility, bliss, and serenity reflect pleasure with calm. Emotional states such as depression, apathy, and dejection reflect distress with calm. The process described above reflects an approach to emotional development known as differential emotions theory. As is evident, this theory hypothesizes that the more subtle, complex emotions are born out of the basic emotional states. Both learning and maturation are involved in emotional development.

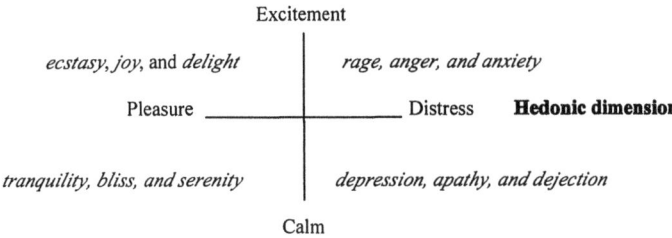

Figure 2.3.1: Emotional States
Source: Bruno, 1992, pp. 105-106

2.3.4 Formation of emotions
2.3.4.1 Mechanisms and phases of emotion formation
- *Mechanisms of emotion formation*

Generally, it may be assumed that human emotions may be induced and modulated in a number of different ways. Pekrun (2000) presents five important mechanisms as the following:

Genetically based emotion An induction of emotions may primarily be based on inherited, hard-wired cognitive schemata, and may be triggered by schema-congruent perceptions. Furthermore, genetical dispositions may predispose an individual to experience specific emotions more or less frequently and more or less intensively, thus contributing to interindividual differences of emotional experiences.

Neurophysilogically mediated emotions Emotions are probably also triggered by basic neurochemical mechanisms (e.g., in psychopathological depressive states).

Conditioned emotions Neuropsychological evidence implies that emotions may be produced by early conditioning establishing direct links between situational perceptions and subcortical, limbic emotional reactions (e.g. LeDoux, 1995).

Cognitively mediated emotions Emotions may be induced by cognitive appraisals of present, past, or future situations.

Habitualized emotions This process of proceduralized emotion occurring basically comprises three stages: (a) automatization of declarative cognitive appraisals; (b) stepwise completion of these appraisals; (c) final short-circuiting of perception and emotion via formation of procedural schemata linking perception and emotion in direct ways.

- *Phases of emotional development*

The three phases of emotional development theory integrates biological, social, and cultural perspectives and explains emotional development with clear defined conceptions. The first phase is *Acquisition*. It includes reflexive affect and temperament as well as acquisition of the labels for emotional categories. One not only acquires and practices different emotions but also demonstrates a style or temperament for them (Haviland, Gebelt, & Stapley, 1997, p.235). Emotions may be temperamental or genetically predisposed (Kagan & Snidman, 1991). The second phase of emotional development is *Refinement*. Refinements of the acquired emotions allow for minimizing expressions, exaggerating them, and covering them with some other signals (deception). In other words, this includes both containment and enhancement strategies that bring emotions into line with social expectations (Haviland et al., 1997, p.238). *Transformation*, the third phase, refers to changes in whole systems. An emotion can remain simple, or it can be transformed to form a system of thoughts, behaviors, and processes. Fear can become subconscious phobia, for example. In this case, tiny bits of information that are traumatic and fearful can set a large system of thoughts and behaviors in motion (Haviland et al., 1997, p.242).

2.3.4.2 Brain development, social development, and a cultural perspective on emotional development

- *Brain development and emotional development*

Affect, an important precursor of cognition Implicit in the developmental model is the idea that during the maturation process, some components of emotional development precede later forms of cognition. As a result, in early development, affect is an important precursor of other modes of thinking, and subsequently must be integrated with other developmental functions for optimal maturation. This appears to be supported by neurobiological study indicating that the emotional/limbic system is operative prior to networks in the brain primary for cognition (e.g., the neocortex) (Greenberg & Snell, 1997). A specific example is infantile amnesia – the inability to regain memories in our conscious mind since early childhood events are stored in the amygdale before language and other representational skills were developed (Jacobs and Nadel, 1985).

The neurophysiological context of emotional development The primary concern in this section is with developmental changes in brain anatomy/neurophysiology and concurrent changes in emotion and emotional regulation. Both the general nature of postnatal brain development and the emerging details regarding the maturation and integration of various structures and pathways have notable implications for emotional development (Sroufe, 1996). Brain growth has been described in terms of stage like processes and qualitative change

(Edelman, 1987; Schore, 1994; Tucker, 1992). It does not develop through the simple addition of new structures, but primarily in terms of changes in the complexity of organization, with increasing integration among components (Sroufe, 1996).

- *Infancy and toddlerhood*

Such a qualitative change would occur in the initial months of life as the cortex first becomes functional. Another is posited at around 9 or 10 months as the frontal lobes mature and the corticolimbic pathways elaborate. Others are posited in the second year as interactive systems come into balance (Schore, 1994). Such coordination between brain development and emotion is not simply because the brain is the physiological substrate for emotion, but because changes in emotional experience in fact affect brain development (Schore, 1994). Communication through affective expression is the major mode of communication available during infancy. By the time the average child has reached two years of age, s/he has an almost complete store of emotional expressions including fear, anger, teasing, enjoyment, persuasion, protest, guilt, anxiety, and affection (Jones, 1992).

- *The preschool years*

Emotional development is accompanied by language development at this stage. With the advent of the child's ability to use self-talk, the control of behavior first comes under conscious cognitive control (Greenberg & Snell, 1997, p.105).

- *The elementary-school-age years*

Between the ages of 5 and 7 children undergo a major developmental transformation that generally includes increase in cognitive processing skills, a growth spurt, and changes in brain size and function (Thatcher, 1994). It is important to recognize that cognitive processes are likely to be effective only if the child has accurately processed the emotional context of a particular situation. Thus, the relationships between affective understanding, cognition, and behavior are of crucial importance in problem solving (Greenberg & Snell, 1997, p.106).

- *Adolescence years*

During the early and middle childhood and adolescence, the child gradually gains self-control over emotional expressions (Jones, 1992). Adolescents begin to become aware of their mood swings and the elicitors of emotions move from family away to peers (Haviland, Gebelt, & Stapley, 1997).

Over the course of childhood, interconnections between the limbic system and neocortex increase and differentiate which allows for the processing of emotional experience to become linked with other areas of the brain and also allows for qualitative changes in emotional development. More specifically, neuronal connections and pathways to and from the limbic system develop. The specific organization of brain organization between the limbic system and both cortical areas and the lower brain stem is unique and will depend on both social experience and genetic factors (Greenberg & Snell, 1997, p.106).

Brain injury and social and emotional development Numerous case studies show that injuries specific to the frontal lobe associate with dramatic changes in emotion regulation and social competence in adulthood (Damasio, 1994). Marlowe (1992) documented some of the most dramatic evidence showing brain-injured children with frontal damage and their socioemotional development deficits. For example, brain-injured children suffered from deficits in self-regulatory behavior and in social awareness, although the children in the care reports often performed in the normal range on tests of cognitive functioning (intelligence testing, reading, mathematics). Their marked deficits in social functioning were observed from childhood throughout the lifespan, primarily in the domains of emotion regulation, adapting to novel situations, and general social behavior.

- *Social development and emotional development*

Figure 2.3.2: Illustration of Bowlby's (1973) Developmental Pathways Concept
Source: Sroufe, L. A. (1996, p. 191)

The progress of emotional development is intertwined with advances in social development. This is not only because the emotions unfold in a social context, but because broader aspects of emotional development, including the regulation of affect, take place within the matrix of care giving relationships (Sroufe, 1989, p. 151), namely in the form of attachment.

One useful model for considering pathways of individual development includes the branching of a tree (Bowlby, 1973, *Figure. 2.3.2*). Any pathway enjoined early (e.g., a pattern of attachment) has multiple possibilities that may be followed based on later contingencies. Two individuals may begin similarly and diverge or begin on different major pathways and ultimately have similar patterns of adaptation due to subsequent turnings in development. The model does imply, however, that early adaptation exercises constraints on later emotional development (Sroufe, 1996).

Greenberg and Snell (1997) stated that in humans, both socioemotional and brain development undergo a protracted course of development, with much room for external influence. Emotional development must be studied in concert with cognitive and social development. This is partly the proposition of holism – that the individual functions as a totality, and no part can be understood in isolation (Gottlieb, 1991; Magnusson, 1988; Werner & Kaplan, 1963). Development is an integrated process, so that other domains of development have profound implications for emotional development, and studying emotional development sheds light on cognitive and social development (Fogel, 1982; Frijda, 1988; Schore, 1994; Thompson, 1990). Current views of the neurobiology of emotional development emphasize the interplay and reciprocal influence that social development and brain development exert on one another. For instance, Panksepp provides examples of developmental milestones in emotion (e.g., social bonding, the development of sexuality during puberty) that may have a bidirectional tie to central nervous system development (Panksepp, 1994).

- *A cultural perspective on emotional development*

Ethological research assumed that there are universal biologically determined program for the expression of emotions (Eid & Diener, 2001). Rather than considering emotions as universal, other psychologists (e.g., Heelas, 1986) view emotions as social constructions which "emphasize aspects that are closely connected with social environment, such as antecedent situations, overt behavior, and culturally specific ways of thinking and talking about emotions" (Mesquita & Frijad, 1992, p.179).

To understand the cultural differences in experiencing of emotions from the psychological perspective, it is important to consider the cultural context (Markus & Kitayama, 1991). In shaping emotional experiences, culture plays a central role (Rosaldo, 1984). Triandis (1997) conceptualized cultural syndrome as "shared set of beliefs, attitudes, norms, values, and behavior organized around a central theme and found among speakers of one language, in one time period, and in one geographic region" (p.443). Dupont concludes that emotional maturity requires people to develop a personal psychology that is congruent with the folk psychology of his or her culture. Personal psychology has meaning only within the context of a culture's folk psychology (Dupont, 1994, p.24).

Idiocentrism and allocentrism Concerning the individual level difference on emotions in different cultures, Eid and Diener (2001) pointed out two reasons with which within cultural differences can be explained. First, the distinction between idiocentrism and allocentrism, which reflect the distinction between individualism and collectivism, and are the concepts, can be used to analyze cultural differences in the individual level. Idiocentrism and allocentrism refer to independent construal of the self and interdependent construal of the self respectively (Markus & Kitayama, 1991). The concept of self is central to an individual's perception, appraisal, and behaviors (Markus & Kitayama, 1991; Triandis, 1989). Culture should be taken into consideration because within the culture context, an individual's concept of self is shaped by cultural norms, values, and beliefs (Shweder & Borune, 1984; Triandis, 1989). The research on cultural influence on self-esteem and embarrassability conducted by Singelis et al. (1995) provides evidence that independent and interdependent self-construes coexist in individuals regardless of culture (see also Gudykunst et al., 1996; Trafimow et al., 1991).

Tight and loose cultures Cultural impact on emotional development can be supported by tight and loose culture reasoning, which is the second reason concerning the individual level difference on emotions in different culture according to Eid and Diener (2001). Triandis (1989) stated that the culture in China is a "collectivistic, but relatively loose country" (p. 511). Loose cultures are characterized as relatively heterogeneous with respect to norms, and show a large variety in norm types; whereas, tight cultures are rather homogeneous with respect to norms and tend to not tolerate deviations. For example, Americans respect more to norms for positive affect, and being unhappy will be regarded as failure. As contrast in China, unhappiness is acceptable because there are Chinese who think positive emotions are undesirable, such as pride. Although pride is not a desirable emotion valued by Chinese people, there are still individuals will view experiencing pride as acceptable (Eid & Diener, 2001).

2.3.4.3 Deterrents to emotional development and individual differences in emotional development

Children growing up in dysfunctional families, children living in the absence of an empathic nurturing relationship, children subject to family abuse, etc., are some of the social experiences believed to be detrimental to children's emotional development. Negative social experiences lead children to develop negative feelings about themselves, significant others, and their lives. They then develop habitual ways of acting on these feelings that are adaptive in their current situations but ineffective and unproductive in the world-at-large. They continue to have mostly negative feelings about themselves and many events and situations, and to have poor interpersonal relationships (Dupont, 1994).

Genetical dispositions may predispose an individual to experience specific emotions more or less frequently and more or less intensively, thus contributing to interindividual differences of emotional experiences (Pekrun, 2000). A general developmental approach was applied to understand individual differences in emotional development by Sroufe (1996). Within broad limits, no fixed amount of tension is associated with negative affect; rather, emotion is dependent on an evaluative process. A major determinant of the infant's affective reaction (and thereby the tendency to maintain organized mastery behavior) is his or her subjective evaluation of an event. In addition to being subject to developmental change, these set points for threat are dramatically influenced by context. As the context is more secure, the threshold for threat is more elevated; more tension may be tolerated. The presence and availability of the infant's caregiver are crucial ingredients in the security of context and have a dramatic influence on the set point for tolerable tension. The infant-caregiver relationship, important in all aspects of emotional development, is thus seen as a cornerstone of individual differences – that is, differences in both tension tolerance and the flexibility of threshold settings (pp. 146-147).

2.3.5 Conceptions of emotionality and emotion regulation

2.3.5.1 A framework of emotionality and regulation

- *Emotion regulation and coping*

Affect regulation involves seeking positive states and avoiding negative states. Positive states are those that enhance or promote one's view of the self, and negative states are those that challenge this view (Markus & Kitayama, 1991). Emotion regulation is defined by Brenner and Salovey (1997) as the process of managing responses that originate within cognitive-experiential, behavioral-expressive, and physiological-biochemical components (p.170). At least three types of regulation processes are relevant to the quality of social functioning: regulation of emotion, regulation of the context itself, and regulation of emotionally driven behavior (Eisenberg, Fabes, & Losoya, 1997, p, 131). Emotions are not seen simply as negative, disruptive outcomes, but as positively motivating and organizing as well. They are integrated with perception and cognition in the service of social and other adaptive behavior (Sroufe, 1996).

The general course of emotional development may be described as movement from dyadic regulation to self-regulation of emotion. Dyadic regulation represents a prototype for self-regulation; the roots of individual differences in the self-regulation of emotion lie within the distinctive patterns of dyadic regulation (Sroufe, 1989, p.151). The caregiver in a sense trains the infant in tension management. This includes the repeated experiences of arousal escalation and de-escalation that occur in dyadic interaction. It also includes the frequent occurrences of distress-relief cycles with which the caregiver is associated (Lamb, 1981). In time, the infant can be more direct and active in seeking what he or she needs by behaving effectively even in the face of high tension (Sroufe, 1996, p.144). A potential for great individual variation may be seen. In contrast to caregivers who finely tune their interactions to move the infant toward ever greater capacity for tension tolerance and self-modulation, other caregiver chronically over stimulate, fail to stimulate, or are strikingly inconsistent in their interactions (Brazelton & Cramer, 1990; Egeland et al, 1993; Fogel, 1993; Schore, 1994; Stern, 1985). Poor quality, anxious attachment will be manifest in dysfunctional dyadic emotional regulation. Children with histories of secure attachment would be predicted to exhibit a notable curiosity, zest for exploration, and affective expressiveness, especially in social situations (Sroufe, 1996).

"Zone of proximal development" (Vygotsky, 1978) and "scaffolding" (Bruner, 1975) refer to the way that more competent others (e.g., parents) can provide frameworks within which children can operate at the limits of their capacities, even stretching them. Thus, within the guidance, boundaries, and support provided by caregivers, children can achieve a substantial level of emotional self-regulation, which prepares the way for the true self-regulation that is to emerge (Sroufe, 1996).

Most of the research views coping as a process comprising two principle components: stressor and strategy (Lazarus & Folkman, 1984). Any effort to manage distress is considered a strategy. Brenner and Salovey (1997, p.170) believe that coping is synonymous with emotion regulation. Successful coping or emotion regulation is determined by the range of strategies available, the ability to select strategies that meet the demands of particular stressors, and the ability to implement these strategies.

Brenner and Salovey's (1997, p.171) review suggests three age-related, developmental trends: (a) Children's use of internal strategies increases throughout development, (b) their use of solitary strategies without parental help increases throughout development, and (c) their ability to distinguish between controllable and uncontrollable stressors, and to effectively match strategy to stressor, improves with development.

- *Internalizing and externalizing behavior*

In research on psychopathology in childhood, two categories of disorders frequently are distinguished: internalizing and externalizing disorders. Internalizing disorders include behaviors that are inner-directed such as anxiety, tendencies to behave in a withdrawn fashion, depressed behaviors, and somatic complaints. External problems, on the other hand, include outwardly expressed behaviors such as hyperactivity, aggression, antisocial behaviors, and conduct problems (Quay, 1986).

The defining characteristics of these problems suggest aspects of emotionality and regulation that are likely to be related to the development and maintenance of specific disorders. For instance, externalizing disorders, which often involve aggression and hostile behavior, suggest under regulation of the experience and expression of anger, a diminished ability to inhibit socially prohibited behavior, and perhaps a lack of fear, which would also serve to inhibit behavior in some situations (Rothbart, Posner, and Hershey, 1995). In contrast, internalizing disorders would appear to involve negative emotionality (e.g., anxiety and despair), lack of adaptive modes of regulation such as attention control, and perhaps extreme levels of behavioral inhibition. However, it is important to keep in mind that internalizing and externalizing symptoms tend to co-occur and usually are substantially correlated; for example, there is a group of children characterized by depression, aggression, and social withdrawal (Garber, et al., 1991).

- *Gender differences in emotionality and emotion regulation*

Perhaps due to gender stereotypes and resulting differential expectations for boys and girls, boys' uninhibited, assertive behavior is viewed more positively than is that of girls. Consistent with this view, inhibited, shy behavior has been associated with positive social interactions and outcomes for girls but not boys (Caspi, et al., 1989). Studies suggest that during middle childhood through adolescence, girls are more likely than boys to rely upon social support strategies to regulate emotion. Second, girls are more likely than boys to use emotion-focused regulation that involves attending to the cognitive-experiential component of emotion (Bull and Drotar, 1991). Finally, research with 8- to 14- year-olds suggests that boys are more likely than girls to use physical exercise to manage distress (Ryan, 1989).

2.3.5.2 Socialization of emotional expression and regulation

Meyer and Salovey (1997) assume that our emotional responses are contextually anchored in social meaning, that is, we have learned cultural messages about the meaning of social transactions, relationships, and even our self-definitions. According to Parke and his colleagues (1989), social environment, parental influence specifically, is critical for children's emotional regulation, in a sense that it influences children's emotionality and emotional regulation directly and indirectly. These two types of socializing influences are distinguished by the parents' intent or objective in socializing the child. Direct socialization influences are intentional attempts by parents to influence or facilitate children's emotional behavior by taking an instructive or organizing role. With indirect socializing influences, parents are not explicitly or intentionally trying to modify their children's emotional behavior. Children's observing the emotional expressions of parents is thought to reflect an indirect influence because parents usually are not trying to influence children's emotional development when they express emotions.

2.3.5.3 Building emotional competence

Mayer and Salovey in their book *Emotional development and emotional intelligence: Educational implications* (1997, p.10) mentioned the concept of emotional intelligence, which includes perceiving, regulating, and thinking about feelings and emotions. Emotional intelligence involves the ability to perceive accurately, appraise, and express emotion; the ability to access and/or generate feelings when they facilitate thought; the ability to understand emotion and emotional knowledge; and the ability to regulate emotions to promote emotional and intellectual growth.

They further associated emotional intelligence with emotional competence, one of the important social competencies. Building emotional competence is an important issue for adolescents. Emotional competence is defined as the demonstration of self-efficacy in emotion-eliciting social transactions (Saarni, 1990). The definition of self-efficacy used here is that the individual has the capacity and skills to achieve a desired outcome (Bandura, 1977). When the notion of self-efficacy is then applied to emotion-eliciting social transactions, it is centered on how people can respond emotionally, yet simultaneously and strategically apply their knowledge about emotions and their emotional expressiveness to relationships with others (Mayer and Salovey, 1997). In this way they can both negotiate their way through interpersonal exchanges and regulate their emotional experiences (Saarni, 1997, p.38). Emotional competence includes the following abilities and capacities (Meyer & Salovey, 1997, pp.47-54):

(1) Awareness of one's emotional state, including the possibility that one is experiencing multiple emotions, and at even more mature levels, awareness that one might also not

be consciously aware of one's feelings due to unconscious dynamics or selective inattention.

(2) Ability to discern others' emotions, based on situational and expressive cues that have some degree of cultural consensus as to their emotional meaning.

(3) Ability to use the vocabulary of emotion and expression terms commonly available in one's (sub)culture and at more mature levels to acquire cultural scripts that link emotion with social roles.

(4) Capacity for empathic and sympathetic involvement in others' emotional experiences.

(5) Ability to realize that an inner emotional state need not correspond to outer expression, both in oneself and in others, and at more mature levels the ability to understand that one's emotional-expressive behavior may impact on another and to take this into account in one's self-presentation strategies.

(6) Capacity for adaptive coping with aversive or distressing emotions by using self-regulatory strategies that ameliorate the intensity or temporal duration of such emotional states (e.g., "stress hardiness").

2.3.6 Relation with needs and values

According to the first postulate of the Neo-Piagetian theory, all of our actions are motivated by our needs and values. One can say that all action...all movement, all thought, or all emotion, responds to a need. Neither the child nor the adult executes any external or internal act unless impelled by a motive; this motive can always be translated into a need (an elementary need, an interest, a question, etc.) (Piaget, 1967, p. 6).

Based on Neo-Piagetian theory (1967), Dupont (1994) elaborates the development of emotions. An emotion begins when some internal or external change puts the system into disequilibrium and a need is created. Dupont further describes the dynamics of need and value. Needs move people to be interested in those things that satisfy the needs, and these interests lead to the creation of values. The emotion ends when the need is met. Piaget (1967) described it this way:" The elementary interests found in children are linked to fundamental organic needs. They are progressively interwoven into complex systems as the child grows up. Much later they will be intellectualized and become scales of value" (p.34).

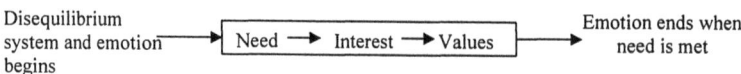

Figure 2.3.3: Neo-Piagetian Theory of Emotional Development

(Adapted from 1967 Neo-Piagetian theory based on Dupont's elaboration, 1994)

Dupont explained further the function of need. As needs are met then, they don't cease to exist; they are still there as background for whatever need is in the foreground. As we mature and have new social experiences, new needs arise and then they reorganize the field. Earlier needs don't cease to exist; they remain as part of the landscape but they are now part of the background, and the field is a new field with the new need now in the foreground. If a need is poorly attended to, it may remain in the foreground and influence the assimilation of all subsequent needs, and, therefore, the organization of the field. Dupont believes that our actions reflect our values, and that what we value reflects our needs (Dupont, 1994, p.29).

2.3.7 Social cognitive control-value theory of achievement emotions

2.3.7.1 Definition and relevance of the domain of achievement emotions

- *Definition of achievement emotions*
Before social cognitive control-value theory of achievement emotions are introduced in details, Schutz and his colleagues' model (*Figure 2.3.4*) of goals as the transactive point for cognition, motivation, and emotions is mentioned as the background supplementary to understanding the concept of achievement emotions.

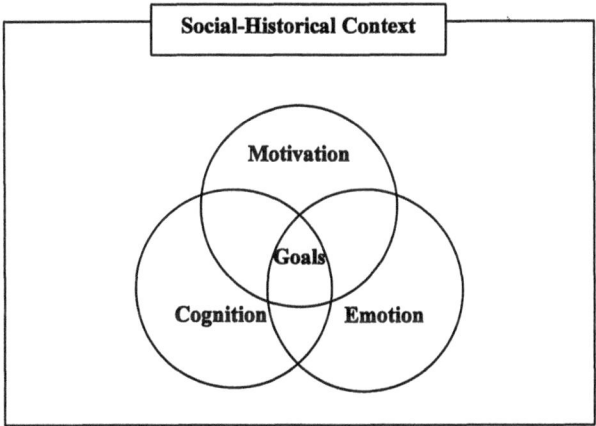

Figure 2.3.4: Representation of Goals as the Transactive Point among Cognition, Motivation, and Emotion (Schutz et al., 2006)

Emotions are ways of being in the world that emerge from appraisals about perceived successes at attaining goals or maintaining standards or beliefs during activities as part of social historical contexts. Emotional experiences in classrooms such as shame, anger and anxiety as well as hope, pride, and joy emerge during transactions that involve the pursuit of various of goals or maintaining various standards (Schutz, Hong, Cross, & Osbon, 2006). Goals are subjective representations (cognition) of what we would like to have happen (motivation) and what we would like to avoid happening in the future (motivation) (Schutz, Crowder, & White, 2001). In the cognitive literature, goals are said to be key organizational processes that influence our thoughts, memories, and interpretations of what we see in the world (Schutz, 1994). In the motivation and self-regulation literature, goals provide direction for our thoughts, behavior, and strategies (Schutz & Davis, 2000). Emotions such as pride, shame, and anxiety are experienced in relation to our goals (Linnenbrink & Pintrich, 2002). Goals, as one of the influencing factors, act as the transactive point for the understanding of constructs such as cognition, motivation, and emotions (Schutz et al., 2006).

Generally emotions may be regarded as systems of interrelated psychological processes including the following (Pekrun, 2000): (a) an activation of subsystems of subcortical brain systems (mainly the limbic system) which is subjectively experienced as emotional feelings (affective component of emotions, e.g., uneasy, nervous feelings in anxiety); (b) emotion-specific cognitions (e.g., threat-related worries in anxiety); (c) emotion-specific motivational tendencies (goals, wishes, and intentions, like wishes to avoid a situation in anxiety); (d) emotion-specific physiological processes (e.g., activation); (e) emotion-specific expressive behavior (cf. Scherer, 1984; Pekrun, 1988). The affective component may be regarded as a necessary defining constituent of the term "emotion".

Table 2.3.1: Classification of Achievement Emotions – Examples

		Positive	Negative
	Concurrent	Enjoyment	Boredom
Self-related	Prospective	Hope for success Anticipatory joy	Anxiety Hopelessness
	Retrospective	Joy after success Relief Pride	Sadness Disappointment Shame/guilt
Social		Gratitude Empathy Admiration Sympathy/love	Anger Envy Contempt Antipathy/hate

Source: Pekrun (2000)

Achievement emotions are defined as emotions tied directly to achievement activities or achievement outcomes (Pekrun, 2006). Activity emotions pertain to ongoing achievement-related activities, and outcome emotions pertain to the outcomes of these activities (Pekrun, 2006; Pekrun, Goetz, Titz, & Perry, 2002a; Pekrun, Elliot, & Maier, 2006). The latter include prospective, anticipatory emotions, as well as retrospective emotions (Pekrun, 2006; see *Table 2.3.1*). As emotions in general, achievement emotions may be classified according to their valence (positive, negative, or neutral), to their contextual frame of reference (individual or social), and to their time reference (concurrent, prospective, or retrospective) (Pekrun, 2000).

Academic emotions are limited to a set of emotions that considered being important in many academic situations and as a consequence may influence students' learning and achievement. As claimed by Pekrun et al. (2002 a), there are three criteria in selecting the set of academic emotions: (1) the emotions belong to the categories of primary human emotions and are important in academic settings; (2) the emotions frequently reported by students which proved by previous studies (e.g., anxiety, enjoyment, hope, pride, anger, boredom, shame) (see Pekrun, 2000); (3) both positive and negative emotions should be included, as well as activating and deactivating emotions. According to Pekrun (2000), joy, hope, relief, pride, anger, anxiety, boredom, and dissatisfaction were frequently experienced by school and university students.

Test anxiety has been researched extensively since the beginning of the 1950s (Mandler & Sarason, 1952; Zeidner, 1998). Test anxiety relates to attention, memory, learning, performance, and achievement motivation across various disciplines, e.g., educational psychology, health psychology, sport psychology, and industrial and organization psychology (Stöber & Pekrun, 2004). Anxiety was the emotion reported most often, accounting for 15% to 25% of all emotions reported in Pekrun et al.'s studies (2002a). In addition, anxiety was mentioned most often not only in relation to taking tests, but also with reference to being in class or studying at home. Although anxiety was the single discrete emotion reported most often, positive emotions were reported no less often than negative emotions, thus pointing to the need to investigate them more thoroughly. However, educational research has paid comparatively little regard to emotions, in particular to positive emotions, except for test anxiety (Pekrun et al., 2002a). Only limited conclusions about a student's emotional experiences in a specific domain can be made from his or her experiences in another subject area, which suggests that it is of critical importance for both teachers and educators to also be aware of the domain specificity of students' emotional experiences (Goetz, Frenzel, Pekrun, & Hall, 2006).

- *Relevance of achievement emotions*

Emotions serve the functions of preparing and sustaining reactions to important events and states by providing motivational and physiological energy, by focusing attention and modulating thinking, and by triggering action-related wishes and intensions. Beyond just being experienced frequently, achievement emotions proved to be relevant for students' motivation, learning strategies, cognitive resources, self-regulation, and academic achievement, as well as to personality and classroom antecedents (Pekrun et al., 2002a).

Findings from Pekrun's studies imply that activating positive emotions like enjoyment of learning and hope for success may exert positive effects on motivation to learn. Deactivating negative emotions like boredom and hopelessness seem to be detrimental for motivation, learning, and achievement. Activating negative emotions like anxiety and anger, however, may exert ambivalent effects (e.g., anxiety may reduce interest and intrinsic motivation, but may at the same time strengthen extrinsic motivation to invest effort in order to avoid failures) (Pekrun, 2000).

2.3.7.2 Assumptions of the theory

- *Emotion formation*

Standards of quality defining achievement are socio-culturally constructed, thus rendering processes of cognitive mediation a central status (Pekrun, 2000, p.148). Pekrun's control-value theory of achievement emotions builds on a more general model of cognitive mediated, reflective types of human motivation and emotion (Pekrun, 1988, 1992c).

- *Control-related and value-related cognitions*

"*Control*" is used here as a generic term denoting causal, functional, and conditional relationships. Control-related cognitions thus refer to subjective appraisals of any type of cause-effect relations, functional relations between variables, or relations between antecedent variables and consequences. Self-concepts of ability, self-efficacy expectancies are examples of control-related cognitions besides causal attributions (Pekrun, 2000, p.149).

Value cognitions according to Pekrun (2000, p.151) refer to intrinsic or extrinsic values of situations, actions, and outcomes for an individual. Intrinsic values refer to inherent properties of the situation, action, or outcome. Extrinsic values pertain to being instrumental for the attainment of other valuable outcomes.

Intrinsic values are closely related to intrinsic goal orientation, whereas extrinsic values are closely related to extrinsic goal orientation. Intrinsic goal orientation is the degree to which students perceive themselves participating in a learning task for reasons such as challenge, curiosity, and mastery (Pintrich, Smith, Garcia, & McKeachie, 1991). An intrinsically motivated student is likely to display autonomy and employ self-exploratory strategies (Bye, Pushkar, & Conway, 2007). By contrast, an extrinsically motivated student seeks approval and external signs of worth and is more likely to ask procedural questions than content-enhancing questions (Sansone & Smith, 2000). In addition to other qualitative differences in learning processes and outcomes, intrinsic motivation has also been consistently distinguished from extrinsic motivation by its association with positive affect (Higgines & Trope, 1990; Kaplan & Maehr, 1999; Ryan & Deci, 2000; Schiefele, 1991).

- *The impact of control and value cognitions on achievement emotions*

Control-value theory postulates that control- and value-related cognitions are of primary importance for achievement emotions (Pekrun, 2000, p.151). Control-value theory is in line with Neo-Piagetian theory of emotional development, which assumes that value induces emotion formation besides other salient factors. More specifically, it may be assumed that any type of achievement emotion depends on value appraisals, and that most of them also depend on control-related cognitions (cf. also Patrick, Skinner, & Connell, 1993). From an educational perspective, appraisals (control beliefs) can be assumed to mediate the impact of situational factors, and can be targeted by educational interventions intended to foster positive emotional development (Pekrun, 2006).

Primary examples for future-related achievement emotions are hope, anticipatory joy, anxiety, and hopelessness relating to future achievement activities and their outcomes. For such emotions to arise, both expectancies and value appraisals of activities/outcomes may be assumed to be necessary. Positive anticipatory emotions may arise when values are positive, negative emotions when values are negative. Examples for retrospective achievement emotions are joy, pride, sadness, and shame. Retrospective emotions are based on recollections of past achievement activities and outcomes. In addition to being remembered, however, these activities/outcomes have to be valued in order to induce emotional feelings, implying that both recollections and value appraisals may be necessary conditions for the induction of such emotions. Enjoyment and boredom are the two most important concurrent achievement emotions. Enjoyment and boredom may be assumed to depend on subjective competences and the perceived amount of control over task performance (Pekrun, 2000, pp. 151-153).

Empirical findings Pekrun (1998, 2000) found that academic self-efficacy, academic control of achievement, and subjective values of learning and achievement related significantly to students' academic emotions. Jacob (1996) investigated test-related achievement emotions among German secondary school students (823 boys, 1044 girls). The test emotions in the study were enjoyment, anger, anxiety, shame and hopelessness. He found that academic self-concept and general self-esteem were positively related to enjoyment of tests, and negatively related to all four negative test emotions. The two value-related variables, on the other hand, positively correlated with both positive and negative test emotions. In line with assumptions of control-value theory, this pattern of correlations implies that control-related cognitions are important for the quality of emotions (positive and negative), whereas high values of achievement may generally lead to increased achievement-related emotionality including both positive and negative emotions. Similar results have been found for achievement emotions in university students (cf. Pekrun & Hofmann, 1999).

He (2004) investigated homework, class, and test-related achievement emotions among Chinese secondary school students (263 male, 282 female, 34 unspecified, mean age = 14.34, SD = 1.01) from diverse social economic backgrounds, and German secondary school students (165 male, 143 female, 4 unspecified, mean age = 14.99, SD = .64), also from diverse social economic backgrounds. The academic emotions in the study were enjoyment, pride, anger, anxiety, shame, hopelessness, and boredom. She found that math self-concept and self-efficacy were positively correlated with enjoyment and pride, negatively correlated with all five negative emotions in the two cultures. The two value-related variables were positively correlated with the both positive emotions and negatively correlated with negative emotions in the two cultures in general.

Ye (2004) investigated class, homework, and test-related emotions among 282 Chinese secondary school students (124 boys, 132 girls, mean age = 15.18, SD = 0.56, gender missing 26) and 312 German secondary school students (165 boys, 143 girls, mean age = 14.91, SD = .64, gender missing 4). The academic emotions in the study were enjoyment, pride, anger, anxiety, shame, hopelessness, and boredom. She found that math self-efficacy was positively associated with enjoyment and pride, negatively correlated with all five negative emotions in the two cultures. The intrinsic value was positively associated with the two positive emotions and negatively associated with negative emotions in the two culture samples, whereas extrinsic value was positively associated with the two positive emotions and negatively associated only with boredom in the two culture samples. Extrinsic value was positively associated with anxiety and shame in the Chinese sample.

Frenzel et al. (2007b) found that girls reported significantly less enjoyment and pride than boys, but more anxiety, hopelessness, and shame, even when controlling for prior achievement. Findings further suggested that the female emotional pattern was due to the girls' low competence beliefs and domain value of mathematics, combined with their high subjective values of achievement in mathematics. Multiple-group comparisons confirmed that the structural relationships between variables were largely invariant across the genders.

- *Proximal social antecedents*

Proximal social antecedents work on individuals' control and value beliefs, which consequently underlie the formation of emotions. The following proximal social antecedents are from Pekrun (2000).

Instruction Individual competences built up by instructional methods used by family, school may underlie the formation of control- and competence- related self-appraisals and, therefore, of achievement emotions. Quality instruction may contribute to positive intrinsic values of achievement activities and, therefore, to the long-term development of activity-related positive emotions. Low-quality instruction may produce boredom.

Task The features of tasks contribute to success and failure, and may therefore be critical for the development of success- and failure-related emotions like prospective enjoyment, anxiety, and hopelessness, or retrospective pride and shame.

Autonomy support versus control On condition that individual capabilities to regulate own activities are sufficiently high, autonomy support can be assumed to be beneficial for the development of competences and competence-related emotions (cf. Deci & Ryan, 1985; Weinert & Helmke, 1995). The Purdue Opinion Poll surveyed 12,000 students attending high schools in various parts of the United States and found that high achievers' families gave their children more autonomy than low achievers' families did; high achievers' families exercised less control on their children than low achievers' families did (Watson & Lindgren, 1979, pp.538-539).

Expectancies and goal structures High expectations may convey messages about high values of achievement, along with restricted definitions of success.

Feedback and consequences of achievement Success and failure feedback may situationally produce achievement-contingent emotions (joy, disappointment, pride, shame, etc.). Beyond situational effects, on condition that feedback is accepted by the individual and attributed accordingly, it may underlie the formation of competence and control appraisals and, therefore, the long-term development of success- and failure-related emotions.

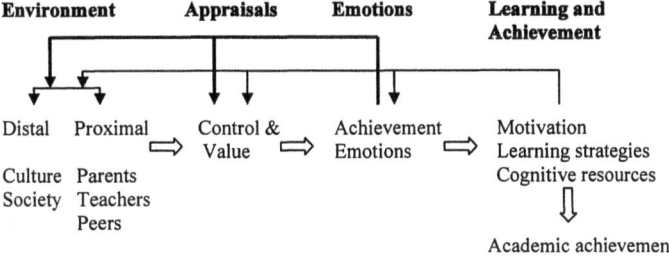

Figure 2.3.5: Social Cognitive Control-Value Theory Model
(Adapted from Pekrun, 2000)

Induction of values Extrinsic and intrinsic values can be socialized. Some important mechanisms may be the following: (a) Information about values may be given directly. (b) Values may be modeled, implying that values conveyed by the behavior of significant others may be adopted. (c) Values may be transmitted by creating learning and work environment which are stimulating.

Using Bronfenbrenner's (1979) terms, proximal social antecedents like those described above are primarily located in the individual's social microsystems. Such microsystems are themselves dependent on subsystems of the society in which they are embedded. These systems themselves are interrelated and dependent on larger systems of cultural norms and values, "the distal environment" in Pekrun's term (2000). The development of achievement emotions may systematically differ between school systems, economical systems, societies, and cultures. Another implication is that differences may be more dependent on immediate environments than on larger systems (Pekrun, 2000, p.159).

Empirical evidence of social factors Numerous studies have found that perceived learning environment is significantly related to student achievement (Fraser, 1994; McRobbie & Fraser, 1993), emotional and social outcomes (Anderman, 2002; Frenzel, Pekrun, & Goetz, 2007a). Jacob (1996) gave some correlation evidence for social antecedents of a number of different achievement emotions in his study on test-related emotions in school students. This study included measures for student perceptions of (a) teacher enthusiasm, (b) pressure for achievement by teachers and parents, (c) positive reinforcement after success, punishment after failure, and support after failure by teachers and parents, and (d) competition within the classroom. In line with the assumptions deduced from control-value theory, teacher enthusiasm, positive reinforcement for success, and support after failure positively correlated with students' enjoyment. Correlations of support with anxiety and other negative emotions were near zero. Punishment after failure, on the other hand, positively correlated with anger, anxiety, shame, and hopelessness. Finally, pressure for achievement and competition within the classroom positively correlate with the negative emotions, but they also tended to related positively to enjoyment. This may be due to the value-inducing functions of achievement expectancies and competitive behavior of significant others which may intensify not only negative achievement emotions, but positive emotions as well.

Goetz, Pekrun, Hall, & Haag (2006) examined student perceptions of their learning environment and emotions experienced within the subject of Latin. In their study, individually perceived positive reinforcement of achievement, teacher enthusiasm, and elaborative instruction in Latin were positively related to individual reports of enjoyment and pride, and negatively related to individual reports of anger and boredom. Achievement pressure from the teacher, however, was positively related to student anxiety and anger, and negatively related to enjoyment and pride in Latin.

Ye (2004) investigated class, homework, and test-related emotions among 282 Chinese secondary school students (124 boys, 132 girls, mean age = 15.18, SD = 0.56, gender missing 26) and 312 German secondary school students (165 boys, 143 girls, mean age = 14.91, SD = 0.64, gender missing 4). This study included measures for student perceptions of (a) teacher enthusiasm, (b) pressure for achievement by parents, (c) parental control and punishment after failure, and (d) peer attitude towards learning mathematics. In line with the assumptions deduced from control-value theory, teacher enthusiasm, and peer attitude were positively correlated with students' enjoyment and pride among the Chinese and German samples and negatively with students' anxiety, anger, hopelessness, and boredom but not shame among the Chinese and German samples. Parental punishment after failure was positively associated with anxiety, anger, hopelessness, boredom, and shame among the Chinese and German samples. Parental pressure for achievement was positively associated with anxiety, anger, hopelessness, and shame but not boredom among the German sample, whereas pressure for achievement was positively associated with enjoyment and pride, and negatively associated with boredom among the Chinese sample.

2.3.8 Hypotheses on achievement emotions

The hypotheses on academic emotions were mainly based on social cognitive control-value theory (Pekrun, 2000). The specific theories on formation of emotions (brain development, social development, cultural impact) and emotion regulation, etc., were reviewed as the broader theoretical background for the specific field of achievement emotions.

The current research focused on Chinese high school students' achievement emotions, namely the intensity of experiencing enjoyment, pride, anger, anxiety, boredom, shame, and hopelessness in math learning context; how their achievement emotions were related to math self-concept, achievement values, achievement goal orientations, and environmental factors; how their achievement emotions display themselves when both achievement and gender were considered. However, the antecedent of collectivistic cultural impact on their achievement emotions was not the focus of the study although cultural context was taken into consideration.

The hypotheses on achievement emotions under the whole framework of the study were projected as the following:

- *Hypothesis 3.1:* *Self-concept and self-efficacy positively correlate with positive emotions and negatively correlate with negative emotions.*
- *Hypothesis 3.2:* *Achievement values correlate with both positive and negative emotions.*
- *Hypothesis 3.3:* *Positive environmental factors positively correlate with positive emotions; negative environmental factors positively correlate with negative emotions.*

There is some interconnection between Arnold's (1960) appraisal theory of emotion and the two general valences of achievement motivation, namely approach and avoidance valences. The two theories, appraisal theory of emotion and achievement motivation theory, give us a lot of insight into attribution, emotion and human motivation. Due to the same valence between positive emotions and approach achievement motivation, and the same valence between negative emotions and avoidance motivation the following hypothesis was made to capture the linkage between the two interrelated concepts and domains.

- *Hypothesis 3.4:* *Approach goals relate positively with positive emotions; avoidance goals relate positively with negative emotions.*

2.4 Attribution theory

2.4.1 Entity and incremental theories of intelligence and attribution theory

Entity (fixed ability) versus incremental (flexible ability) theories of intelligence (Dweck, 2000) have been discussed in regard to their relationships with educational constructs such as attribution theory (e.g., Hong, Chiu, Dweck, Lin, & Wan, 1999; Stipek & Gralinski, 1996). When faced with unsatisfactory performance, incremental theorists tend to put forth more efforts to take remedial action to improve. On the other hand, entity theorists (students holding a fixed ability) are more concerned with demonstrating their ability or not revealing their lack of it, and they tend to attribute their failure to lack of ability, rather than effort (Hong et al.,

1999). Studies with gifted students have indicated that these students tend to hold an incremental view, rather than an entity view of ability (Feldhusen & Dai, 1997; Hsueh, 1997).

2.4.2 Development of causal structure

- *Heider (1958) and locus dimension*

The first systematic analysis of causal structure was proposed by Heider (1958). Rightly called the originator of the attributional approach in psychology, Fritz Heider has been in the background of much of the present theory (Weiner, 1985). The most fundamental causal distinction made by Heider (1958) is factors within the person and factors within the environment (p.82). Since the early 1950s, psychologists have adopted an internal-external distinction (see Collins, Martin, Ashmore, & Ross, 1974). But the domination of internal-external comparisons in psychology arrived with the work of Rotter (1966), for his classification of individuals into internals and externals became a focus for research. Thus, the analysis of the structure of causality logically began with an internal-external (locus) dimension (Weiner, 1985).

- *Weiner (1971) and locus of causality, and stability*

Weiner et al. (1971) further developed attribution theory, including two dimensions of causality, namely, locus of causality and stability. The most dominant causes in achievement-related contexts are ability, effort, task difficulty, and luck, which are within a 2 times 2 categorization scheme. Ability was classified as internal and stable, effort as internal and unstable, task difficulty was thought to be external and stable, and luck was considered external and unstable. However, Weiner (1983) did not believe that the four cells truly represented the classification system. Ability may be perceived as unstable if learning is possible; effort often is perceived as a stable trait, captured with the labels of lazy and industrious; tasks can be changed to be more or less difficult; and luck may be thought of as a property of a person (lucky or unlucky).

- *Weiner (1979) and locus of causality, stability, and controllability*

The deductive reasoning that led to the identification of the third property of causality was based on Rosenbaum (1972). Rosenbaum recognized that mood, fatigue, and temporary effort, all are internal and unstable causes. Yet they are distinguishable in that effort is subject to volitional control – an individual can increase or decrease effort expenditure. Mood or fatigue, in contrast, cannot be willed to change. Therefore, controllability was identified as the third property of causality (Weiner, 1979).

2.4.3 An attribution theory of achievement motivation and emotion

Weiner incorporated a full range of cognitions and emotions in his attribution theory, and there is an explicit concern with the self. Furthermore, an attempt has been made to relate the structure of thought (in this case, causal thinking) to the dynamics of feeling and action. Weiner proposed a theory of motivation and emotion in which causal ascriptions play a key role. There are a few dominant causal perceptions in achievement-related contexts. The perceived causes of success and failure share three common properties: locus, stability, and controllability (1985). The perceived stability of causes influences changes in expectancy of success; all three dimensions of causality affect a variety of common emotional experiences, including anger, gratitude, guilt, hopelessness, pity, pride, and shame. Guided by expectancy and value theory, Weiner presumed expectancy and affect, in turn, direct motivated behavior. The theory therefore relates the structure of thinking to the dynamics of feeling and action (Weiner, 1985).

The child who holds himself responsible – whose locus of control is internal – is more likely to attempt to solve problems and strive toward distant goals than are children who believe there is little they can do to affect the events in their life. Such children tend to see their locus of control as external (Watson & Lindgren, 1979, p.372). In the summer school evaluation program (Neber & Heller, 2002), gifted students were found to attribute success to their internal variable and controllable factors (effort) or stable causes (ability); in explaining possible failures, gifted students were found to attribute controllable factors either referring to external causes (task difficulty) or variable internal factors (effort). These tendencies correspond to other findings with gifted groups of students, whereas nongifted and low achieving students overestimate ability as a cause of failures (Dai, Moon, & Feldhusen, 1998).

Kelley (1971) stated why people are attributors, and continuously attribute reasons for their outcome of behavior. It clearly is functional to know why an event has occurred. "The attributor is not simply an attributor, a seeker after knowledge; his latent goal in attaining knowledge is that of effective management of himself and his environment." (Kelley, 1971) Once a cause, or causes, is assigned, effective management may be possible and a prescription or guide for future action can be suggested. If the prior outcome was a success, then there is likely to be an attempt to reinstate the prior causal network. On the other hand, if the prior outcome or event was undesired – such as exam failure – then there is a strong possibility that there will be an attempt to alter the causes to produce a different (more positive) effect (Weiner, 1985).

2.4.4 Motivational dynamics of perceived causality
2.4.4.1 Expectancy change

Weiner constructed an attributional theory of motivation, which has some linkage between attributional thinking and goal expectancy. A fundamental psychological law relating perceived causal stability to expectancy change was documented by Weiner (1985). Weiner's expectancy principle states that changes in expectancy of success following an outcome are influenced by the perceived stability of the cause of the event. The principle has three corollaries:

Corollary 1 If the outcome of an event is ascribed to a stable cause, then that outcome will be anticipated with increased certainty, or with an increased expectancy, in the future.

Corollary 2 If the outcome of an event is ascribed to an unstable cause, then the certainty or expectancy of that outcome may be unchanged or the future may be anticipated to be different from the past.

Corollary 3 Outcome ascribed to stable causes will be anticipated to be repeated in the future with a greater degree of certainty than are outcomes ascribed to unstable causes.

2.4.4.2 Affective reactions

- *The attribution – emotion process*

Weiner (1985) offered an attributional view of the emotion processes and tried to document pattern linking attributional thinking and specific feelings. The attributional framework assumes (Weiner, 1985) a sequence in which cognitions of increasing complexity enter into the emotion process to further refine and differentiate experience. In another word, increasing cognitive involvement generates more differentiated emotional experience (Abelson, 1983; Roseman, 1984; C. Smith & Ellsworth, 1985).

Outcome evaluation Weiner (1985) stated that following the outcome of an event, there is a general positive or negative reaction (a "primitive" emotion) based on the perceived success or failure of the outcome (the "primary appraisal"). These emotions are labeled as outcome dependent - attribution independent.

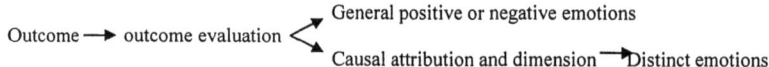

Figure 2.4.1: The cognition – emotion process
(Source: Weiner, 1985)

Past successes and failures themselves have been shown to elicit characteristic affective responses (Weiner, Russell, and Lerman, 1978). Success, especially on challenging tasks, leads to positive feelings; failure, especially on easy tasks, leads to negative feelings (Harter, 1980). Other things being equal, these affective responses should influence the enjoyment or intrinsic value of subsequent related activities (Bandura, 1977). One should like activities that have been associated with positive feelings in the past more than activities that have been associated with negative feelings (Eccles, 1983).

Causal attribution Following outcome appraisal and the immediate affective reaction, a causal ascription will be sought. A different set of emotions is then generated by the chosen attribution(s). Emotions which are dependent on causal attributions are labeled attribution dependent. For example, failure perceived as due to bad teaching quality produces anger; failure perceived as due to lack of effort produces shame. Additionally, causal dimensions play a key role in the emotion process. Each dimension is uniquely related to a set of feelings. For example, success and failure perceived as due to internal causes such as personality, ability, or effort respectively raises or lowers self-esteem or self-worth, whereas external attributions for positive or negative outcomes do not influence feelings about the self. Hence, self-related emotions are influenced by the causal property of locus, rather than by a specific cause per se (Weiner, 1985).

Weiner's cognition-emotion process model is very similar with Arnold's (1960) appraisal theory of emotion, which builds the groundwork for the sequences of emotion formation. Arnold's position, which closely resembles that of Aristotle, views the emotion process as a sequence of events that begins with the perception of the stimulus, immediately followed by an act of appraisal, which in turn activates the emotional response. She defines perception as the immediate, direct apprehension of the stimulus, and appraisal as a judgment of it as either good or bad. It is the appraisal of the stimulus in relation to the perceiver that triggers a feeling toward the object, person, or event that, in turn, produces a behavioral response of approach (if the object is judged to be likely to benefit the self) or avoidance (if the object is appraised as potentially harmful). Neutral stimuli are simply ignored. Arnold extends Aristotle's view that emotion can in turn influence cognition, by noting that emotions can organize and bias later perceptions. This process can lead to situations in which the perception-appraisal of an object, person, or event can stabilize and generalize to a whole class of people, objects, or events (LaFreniere, 2000).

- *Dimension-related emotions*

The emotion of pride and feelings of self-esteem are linked with the locus dimension of causality; anger, gratitude, guilt, pity, and shame all are connected with the controllability dimension; and feelings of hopelessness (hopefulness) are associated with causal stability. Weiner, Russell, and Lerman (1978, 1979) have provided empirical support for the link between attributions and affective responses. Weiner et al. (1978) have found that attributing one's success internally leads to feelings of pride, satisfaction, and competence, while attributing success externally leads to feelings of gratitude and surprise. Attributing one's failure to internal causes leads to feelings of guilt, resignation, and regret, while attributing failure to external causes leads to feelings of anger and surprise. Jones (1992) also found that success due to internal causes such as ability and effort produces pride and positive self-esteem; failure due to low ability results in humiliation, and failure due to lack of effort produces guilt. Thus, it appears that attributions influence, in part, the affective responses one experiences in achievement setting, which, in turn, should influence the value of these tasks (Eccles, 1983).

Pride (self-esteem) It is reasoned that pride and positive self-esteem are experienced as a consequence of attributing a positive outcome to the self and that negative self-esteem is experienced when a negative outcome is ascribed to the self (Stipek, 1983; Weiner et al., 1978, 1979). As Harvey and Weary (1981) noted, "By taking credit for good acts and denying blame for bad outcomes, the individual presumably may be able to enhance or protect his or her self-esteem" (p. 33). Pride and personal esteem therefore are self-reflective emotions, linked with the locus dimension of causality (Weiner, 1985).

Anger More than anything else, anger is an attribution of blame (Averill, 1983, p.1150). The attributional antecedent for anger is an ascription of a negative, self-related outcome or event to factors controllable by others (Weiner, 1980a, 1980b; Weiner, Graham, & Chandler, 1982).

Pity In contrast to the linkage between controllability and anger, it is hypothesized that uncontrollable causes are associated with pity (Weiner, 1985).

Guilt and shame Guilt is elicited by controllable causes and usually directed inward. Shame is elicited by self-related and uncontrollable causes, such as lack of ability (Weiner, 1985). In studies testing uncontrollability-shame and controllability-guilt associations, Brown and Weiner (1984), Covington and Omelich (1984), and Jagacinski and Nicholls (1984) have reported that shame-related affects (disgrace, embarrassment, humiliation, and/or shame) are linked with failure due to low ability, whereas guilt-related affects (guilt, regret, and/or remorse) are associated with failure due to lack of effort. It has been also documented that shame-related emotions give rise to withdrawal and motivational inhibition, whereas guilt-related emotions promote approach behavior, retribution, and motivational activation (Hoffman, 1982; Wicker et al., 1983). Hence there are linkages between low-ability-shame-inhibition and between lack-of-effort-guilt-augmentation (Weiner, 1985).

Hopelessness It has been found that hopelessness is elicited given an attribution for a negative outcome to stable causes (Weiner et al., 1978, 1979).

2.4.5 Chinese cultural influences on achievement attributions

Studies have revealed that causal attributions and their dimensional meaning may be mediated by sociocultural factors (cf. Betancourt & Weiner, 1982; Chandler, Shama & Wolf, 1983; Fry & Ghosh, 1980; Hess, Chang, & McDevitt, 1987; Salili, Hwang & Choi, 1989). Several studies have reported strong cultural differences between Chinese and western students in ability and effort attributions (cf. Hau & Salili, 1990, 1991; Holloway, 1988). In the West, ability to achieve is an important human value. Ability is considered to be a given and stable attribute in the West (Weiner, 1986). Chinese, on the other hand, place great emphasis on effort and tend to attribute both success and failure more to internal and controllable causes such as effort and study skills than to ability (Hau & Salili, 1989; Salili, Hwang & Choi, 1989). This reflects the importance of effort and endurance in the Chinese culture. Achievement through effort and hard work is more highly valued than achievement through ability. Ability is considered less important and modifiable through effort (Salili, 1995). Consistent with Chinese cultural values, Chinese teachers seldom praise their students and consider it harmful to the development of the child's character if it is given frequently and without an outstanding cause. On the other hand, Chinese teachers do not hesitate to administer punishment, severe discipline is often considered useful and necessary in children's education. Therefore, students learn from an early age to work hard but not to expect any external reward (Salili, 1995).

2.4.6 Achievement change programs

The main empirical finding in these studies is that persistence in the face of failure is enhanced when attributions for failure are changed from low ability to lack of effort (Andrews & Debus, 1978; Chapin & Dyck, 1976; Dweck, 1975; Zoeller, Mahoney, & Weiner, 1983), to poor strategy (Anderson, 1983b; Anderson & Jennings, 1980), or to temporary external barriers (Wilson & Linville, 1982, 1985). However, ascriptions to ability, effort, strategy, and external barriers have different affective consequences. Shifting ascriptions of failure due to ability to lack of effort is altering reactions of shame to guilt. On the other hand, a program that promotes task difficulty ascriptions (Wilson & Linville, 1982, 1985) theoretically is enhancing the self-esteem of the participants. Perhaps increments in self-esteem rather than (in addition to) expectancy maintenance are responsible for the augmented achievement strivings (Weiner, 1985).

2.4.7 Hypotheses on attributions

The current research focused on Chinese high school students' attribution reasoning in success and failure contexts, namely whether they attributed success to ability, effort, task difficulty, luck, or learning strategy use; whether they attributed failure to ability, effort, task difficulty, luck, or learning strategy use; how their attributions were related to achievement emotions. However, the collectivistic cultural impact on attributions was not the focus of the study although cultural context was taken into consideration. The hypotheses on attributions under the whole framework of the study were projected as the following:

- *Hypothesis 4.1:* Attributional antecedents for pride are internal causes, e.g., ability and/or effort.
- *Hypothesis 4.2:* Attributional antecedents for anger are controllable causes, e.g., effort, learning strategy use, or quality of instruction.
- *Hypothesis 4.3:* Attributional antecedents for shame are internal and uncontrollable causes, e.g., lack of ability.
- *Hypothesis 4.4:* Attributional antecedents for hopelessness are stable internal causes, e.g., lack of ability.

2.5 Self-regulated learning

Teaching students to be life-long learners is an important goal of higher education. Students need to be taught explicitly how to use learning strategies to achieve this goal (Pintrich,

McKeachie, & Lin, 1987). Learning strategies is one of the components of self-regulated learning. Self-regulated learning, according to Pintrich (1988, 1989, 1990), includes two general components, namely, a self-regulated learning component and a motivational component.

2.5.1 Definitions of self-regulated learning and learning strategies

Self-regulation of cognition and behavior is an important aspect of student learning and academic performance in the classroom context (Corno & Mandinach, 1983; Corno & Rohrkemper, 1985). The working definition of self-regulated learning was summarized by Pintrich et al. (1990) to include the following three key components. First, self-regulated learning includes students' metacognitive strategies for planning, monitoring, and modifying their cognition (e.g., Brown, Bransford, Campione, & Ferrara, 1983; Corno, 1986; Zimmerman & Pons, 1986, 1988). Students' management and control of their effort on classroom academic tasks has been proposed as another important component. A third important aspect of self-regulated learning that some researchers have included in their conceptualization is the actual cognitive strategies that students use to learn, remember, and understand the material (Corno & Mandinach, 1983; Zimmerman & Pons, 1986, 1988). Weinstein and Mayer (1986) defined learning strategies to include any cognitions and behaviors that influence the encoding process and facilitate acquisition and retrieval of information. Different cognitive strategies such as rehearsal, elaboration, and organizational strategies have been found to foster active cognitive engagement in learning and result in higher levels of achievement.

According to Pintrich, Smith, Garcia, & McKeachie (1991), cognitive and metacognitive learning strategies are made up of five subscales, namely, rehearsal, elaboration, organization, critical thinking, and metacognitive self-regulation.

Basic cognitive strategy – rehearsal Basic rehearsal strategies involve reciting or naming items from a list to be learned. These strategies are assumed to influence the attention and encoding processes, but they do not appear to help students construct internal connections among the information or integrate the information with prior knowledge.

Complex cognitive strategy – elaboration Elaboration strategies help students store information into long-term memory by building internal connections between items to be learned. Elaboration strategies include paraphrasing, summarizing, creating analogies, and generative note-taking. These help the learner integrate and connect new information with prior knowledge.

Complex cognitive strategy – organization Organization strategies help the learner select appropriate information and also construct connections among the information to be learned. Examples of organizing strategies are clustering, outlining, and selecting the main idea in reading passages. Organizing is an active, effortful endeavor, and results in the learner being closely involved in the task.

Critical thinking Critical thinking refers to the degree to which students report applying previous knowledge to new situations in order to solve problems, reach decisions, or make critical evaluations with respect to standards of excellence.

Metacognitive self-regulation Metacognition refers to the awareness, knowledge, and control of cognition. There are three general processes making up metacognitive self-regulatory activities: planning, monitoring, and regulating. Planning activities refer to goal setting which helps to activate relevant aspects of prior knowledge that make organizing and comprehending the material easier. Monitoring activities include tracking of one's attention as one reads, and self-testing and questioning. Regulating refers to the fine-tuning and continuous adjustment of one's cognitive activities Regulating activities are assumed to improve performance by assisting learners in checking and correcting their behavior as they proceed on a task (Pintrich et al., 1991).

2.5.2 Motivational beliefs in learning

Knowledge of cognitive and metacognitive strategies is usually not enough to promote student achievement; students also must be motivated to use the strategies as well as regulate their cognition and effort (Paris, Lipson, & Wixson, 1983; Pintrich, 1988, 1989; Pintrich, Cross, Kozma, & McKeachie, 1986). There also is evidence to suggest that students' perceptions of the classroom as well as their individual motivational orientations and beliefs about learning are relevant to cognitive engagement and classroom performance (e.g., Ames & Archer, 1988; Nolen, 1988). Accordingly, it is important to examine how the three components of self-regulated learning are linked to individual differences in student motivation in order to describe and understand how personal characteristics are related to students' cognitive engagement and classroom academic performance (Corno & Snow, 1986; Snow, 1989; Weinert, 1987).

The theoretical framework for conceptualizing student motivation is an adaptation of a general expectancy-value model of motivation (cf., Eccles, 1983; Pintrich, 1988, 1989). The model proposes that there are three motivational components that may be linked to the three different components of self-regulated learning: (a) an expectancy component, which includes students' beliefs about their ability to perform a task, (b) a value component, which includes students' goals and beliefs about the importance and interest of the task, and (c) an affective component, which includes students' emotional reactions to the task.

2.5.2.1 Expectancy component

The expectancy component of student motivation has been conceptualized in a variety of ways in the motivational literature (e.g., perceived competence, self-efficacy, attributional style, and control beliefs), but the basic construct involves students' beliefs that they are able to perform the task and that they are responsible for their own performance. In general, the research suggests that students who believe they are capable engage in more metacognition, use more cognitive strategies, and are more likely to persist at a task than students who do not believe they can perform the task (e.g., Fincham & Cain, 1986; Paris & Oka, 1986; Schunk, 1985). Hofer and Yu (2003) found that self-efficacy positively correlated with cognitive strategies use. Self-efficacy beliefs contribute to motivation by determining the goals people set for themselves, how much effort they expend, and how long they persevere (Bandura, 1986, 1993; Schunk, 1987).

2.5.2.2 Value component

The value component of student motivation involves students' goals for the task and their beliefs about the importance and interest of the task. Although this component has been conceptualized in a variety of ways (e.g., learning vs. performance goals, intrinsic vs. extrinsic orientation, task value, and intrinsic interest), this motivational component essentially concerns students' reasons for doing a task. The research suggests that students with a motivational orientation involving goals of mastery, learning, and challenge, as well as beliefs that the task is interesting and important, will engage in more metacognitive activity, more cognitive strategy use, and more effective effort management (e.g., Ames & Archer, 1988; Dweck & Elliott, 1983; Eccles, 1983; Meece, Blumenfeld, & Hoyle, 1988; Nolen, 1988; Paris & Oka, 1986). Hofer and Yu (2003) found that achievement values positively correlated with cognitive and metacognitive strategies use.

2.5.2.3 Affective component

The third motivational component concerns students' affective or emotional reactions to the task. Again, there are a variety of affective reactions that might be relevant (e.g., anger, pride, guilt), but in a school learning context one of the most important seems to be test anxiety (Wigfield & Eccles, 1989). Research on test anxiety has been linked to students' metacognition, cognitive strategy use, and effort management (e.g., Benjamin, McKeachie, Lin, & Holinger, 1981; Culler & Holahan, 1980; Tobias, 1985). The cognitive models of test anxiety (e.g., Benjamin, McKeachie, & Lin, 1987; Tobias, 1985) propose that for some test-anxious students who actually have adequate cognitive skills, test anxiety during exams engenders worry about their capabilities that interferes with effective performance.

Although the other two motivational components generally show simple, positive, and linear relations with the components of self-regulated learning, the results for test anxiety are not as straightforward. For example, Benjamin et al. (1981) found that although high-anxious students seemed to be as effortful and persistent as low-anxious students, they appeared to be very ineffective and inefficient learners who often did not use appropriate cognitive strategies for achievement. Furthermore, other research suggests that high-anxious children are not persistent or avoid difficult tasks (Hill & Wigfield, 1984). Accordingly, test anxiety may be related to the three components of self-regulated learning in different ways.

Previous research suggests that the expectancy and value components will be positively related to the three self-regulated learning components, whereas the research on test anxiety does not suggest such simple relations (Pintrich & De Groot, 1990). Pintrich and De Groot (1990) found that expectancy and value components positively related to strategy use and self-regulation, whereas test anxiety did not relate to strategy use or self-regulation significantly. However, Hofer and Yu (2003) found that test-anxiety negatively correlated with cognitive and metacognitive strategies use. Concerning motivation and self-regulated learning and performance, Pintrich and De Groot (1990) found that higher levels of intrinsic value and self-efficacy were associated with higher levels of student achievement; higher levels of test anxiety were only significantly related to lower levels of performance on exams; higher levels of cognitive strategy use and self-regulation were associated with higher levels of achievement. Eaton and Dembo (1997) found that Asian-American students reported significantly higher levels of text anxiety than Euro-American students, and argued that Chinese students may be motivated more by a fear of failure than Euro-American students. Rao and Sachs (1999) further argued that test anxiety may have a facilitative influence on Chinese students' academic achievement.

Teaching students about different cognitive and self-regulatory strategies may be more important for improving actual performance on classroom academic tasks, but that improving students' self-efficacy beliefs may lead to more use of these cognitive strategies (cf., Borkowski, Weyhing, & Carr, 1988; Garner & Alexander, 1989; Schunk, 1985). Furthermore, intrinsic value did not have a significant direct relation to student performance in any of the regressions that included cognitive strategy use or self-regulation. The cognitive variables, self-regulation in particular, were better predictors of actual academic performance (Pintrich & De Groot, 1990). This finding parallels the work of Eccles (1983), who found that value components did not have a direct influence on student achievement in math but were closely tied to students' choice of future math courses.

2.5.3 Hypotheses on learning strategy use

The current research focused on Chinese high school students' learning strategy use, namely the intensity of applying rehearsal, elaboration, organization, critical thinking, and metacognitive strategies in mathematics learning process; how their learning strategy use was related to math self-concept, achievement values, and achievement emotions. However, the collectivistic cultural impact on learning strategy use was not the focus of the study although cultural context was taken into consideration. The hypotheses on learning strategies under the whole framework of the study were projected as the following:
- *Hypothesis 5.1:* Math self-concept and Self-efficacy positively correlate with learning strategies use.
- *Hypothesis 5.2:* Achievement values positively correlate with learning strategy use.

- **_Hypothesis 5.3:_** *Positive emotions positively correlate with learning strategy use; negative emotions negatively correlate with strategy use.*

2.6 Translation of measures

The measures applied in the current study were translated from English into Chinese. Translation of a source language into a target language is a crucial undertaking for anyone interested in cross-cultural research (Werner & Campbell, 1970). Translation of instruments must therefore be properly handled.

2.6.1 Perspectives

The Emic-Etic distinction   The concepts "emic" and "etic" refer to different approaches to understanding culture. Emic studies involve monocultural approaches, or those involved with studying the culture on its own terms and employing culture-based constructs. Etic research, on the other hand, seeks to understand or to compare cultures using a common framework for each. However a pseudoetic approach is unfortunately taken when a procedure or a test developed in a Western culture is applied to another group that is highly different in tradition, language, cultures, or background (Butcher, 1982).

From one culture to another   The measures applied in the present study were Pintrich and his colleagues' Motivated Strategies for Learning Questionnaire, Pekrun and his colleagues' Academic Emotions Questionnaire for Mathematics, etc., which are all in English and therefore needed to be translated into the target language of Chinese. The English AEQM was reliably translated from German into English by Pekrun and his colleagues. One of the earliest cross-cultural clinical ventures involved the use of psychological tests developed in one culture to understand the cognitive behavior or personality of individuals in a different context. Present-day test translation and adaptation studies are aimed at providing a workable, valid alternate form of the test for use in the new culture and the application of the test's constructs and interpretations for individual personality description and clinical prediction in the new culture (Butcher, 1982).. For example, the MMPI has been widely translated and adapted for clinical use in over 45 countries (Butcher & Pancheri, 1976; Butcher & Clark, 1979).

One important requirement for test equivalence is that tests used in cultural contexts that are different from their original environment show a similar or equivalent factor structure in the target culture. Linguistic equivalence, which is the accurate translation of instructions, questionnaires, and other material, is consequently an important task. Materials must communicate the appropriate psychological meanings as well as being linguistically accurate. Although many technical problems and some alterations are frequently required, especially in countries with wide cultural differences, many of the translated methods and instruments thrive in the new cultures (Butcher, 1982).

Decentering On the one hand, there is symmetrical or decentered translation aiming at both loyalty of meaning and equal familiarity and colloqialness in each language. On the other hand, there is asymmetrical or unicentered translation, in which loyalty to one language, usually the source language, dominates. The symmetrical translation is an ideal method for much literary translation. Decentering implies a de-emphasis of the investigator's language in such a way that the system of symbols supersedes a single culture. At best decentering eliminates the distinction between source and target language and stresses equivalences (Werner & Campbell, 398-399, 1970). Also according to them, multiple-stage translation and back-translation are some of the methods for decentering. Back translation provides a most useful technique for suggesting revisions of the first translation effort.

2.6.2 Procedures

Cross-cultural research using psychological instruments has been greatly hampered by the concept of "standardized tests, "i.e., the belief that there is well-developed all-purpose comparison instruments for which valuable "population norms" have been collected. This notion of "standardized tests", the almost magical assumption that psychological tests could measure directly and perfectly what they claimed to measure, has led to a most unfortunate ethnocentric asymmetry in the translation of such instruments. A fundamental asymmetry has resulted in which familiar, colloquial, accessible test items in English become exotic, awkward, and difficult items in the target language (Werner & Campbell, 1970).

Ideally, the method of treating the original-language version as itself up for revision should be adopted. Such revision and item-selection processes will no doubt produce an original version which is more banal, less subtle, more explicit, less colloquial, less idiomatic, less metaphorical, and with less extreme items. This may produce a test which is less reliable and less valid for original-language use. It will require collecting new original-language comparison groups on the new instrument (Schachter 1954). Practically, the method of adapting the items of the original test to the target culture should be applied.

Butcher (1982) gave some suggestions how a translation of instruments could be prepared.

(1) Using more than one translator, working independently or in committees to prepare an initial translation of the material
(2) Using a back translation procedure to detect any difficulties or failed communication
(3) Field testing translated material – assuring readability, style, and accuracy of content communication
(4) Conducting bilingual studies in the case of translated tests
(5) And determining the psychometric adequacy or translanguage equivalence of the test in all the target languages.

Seemingly, Campbell et al., (1970) pointed out that a researcher might use one or more of the following techniques: back-translation, bilingual technique, committee approach, and pretest procedures. A similar seven-step procedure was recommended by Brislin (1970), which provides some guidelines on preparing translations that aim to communicate clearly both the linguistic and psychological meaning of the material. The common approaches in the above mentioned translation procedures are using bilinguals, back translation, committee approach, and field test.

Schachter (1954) believes that a suggested tactic for confirming the equivalence of translation is to have bilinguals respond to the items in both languages. Used as a formal statistical approach, the goal should not be identity at the item-by-item level, but rather equivalence of means and variances, plus appropriate correlations between scores on the two forms. Most commonly, claims for adequacy of test-translation are based upon finding similarity in reliability, validity, factor structure and other statistics for the forms in both languages.

Chapter III METHODOLOGY

3.1 Design of the study

The current study applied cross-sectional method to describe the associations among affective, cognitive, and motivational variables. Correlation and regression statistics were the main means to achieve this goal. Sophisticated and widely recognized questionnaires were applied to collect data and pin-point affective-cognitive learning processes and outcome pattern.

3.2 Sample characteristics

3.2.1 Sample size and gender

Table 3.2.1: Descriptive Statistics of Gender, Sample Size, and School Population

School	Male	Female	Gender missing	Sample size	School population	Sample school population ratio
Elite A	138	92	1	231	2100	11%
Ordinary B + C	97	132	4	233	4000	5.8%
Ordinary B	44	98	2	144	1500	9.6%
Ordinary C	53	34	2	89	2500	3.6%
Whole sample	235	224	5	464	6100	7.6%

Note.
A: The elite senior high school
B: One of the two ordinary senior high schools
C: One of the two ordinary senior high schools
B+C: Sample and population descriptive statistics combined out of the two ordinary senior high schools

The sample for the study were 464 senior high school students from two metropolitans in China, one in north China of Hebei province and the other in middle China of Hubei province. Both of the two cities have more than 5 million people. The 464 senior high school students were from three schools, one elite high school, and two ordinary high schools. The elite high school sample had 231 students, among them 138 males and 92 females. One participant from the elite sample did not reveal his or her gender. The elite high school had a student size of about 2100. About 11 % students of elite high school took part in the study. The ordinary high school sample had 233 students, among them 97 males and 132 females. Four participants from the ordinary sample did not reveal their gender. The two ordinary high schools had a student size of about 1500 and 2500, respectively. About 9.6 % and 3.6 % students of the two ordinary high schools took part in the study.

3.2.2 Social characteristics of the sample

- *Sample age*

The mean age for the elite high school sample was 17.36 (SD = .52). The mean age for the ordinary high school sample was 16.96 (SD = .53). Grand total mean age was 17.16 (SD = .56).

Table 3.2.2: Socioeconomic Characteristics of the Sample

Characteristic	High achievement group Mean	SD	Low achievement group Mean	SD	Grand total Mean	N	SD	F
Child's age	17.36	.52	16.96	.53	17.16	457	.56	64.78***
Math GPA	126.73	12.77	87.60	21.80	108.52	363	26.26	448.47***
Mother's education level	3.47	.93	3.12	.87	3.30	455	.91	17.04***
Father's education level	3.70	.87	3.30	.86	3.50	455	.89	24.71***
SES	2.81	.79	2.58	.81	2.70	459	.81	9.73**

Note. **p < .01 ***p < .001
Education level 1. Primary school   2. Junior high school   3. Senior high school   4. Bachelor
                 5. Master   6. Doctor   7. Post-doctor
SES 1. Not at all wealthy 2. Not wealthy 3. Average   4. Relatively wealthy   5. Very wealthy

- *Math GPA (Math grade point average)*

Math GPA was the average of the last mid-term mathematics exam score and the last final-term mathematics exam score. Since the mid-term and final-term mathematics exams took place 4 and 2 months respectively prior to the assessment of beliefs and emotions, indicating that there was a temporal space between the assessment of prior achievement level and beliefs and emotions.

Math GPA of high achievement sample was 126.73 (SD = 12.77). Math GPA of low achievement sample was 87.60 (SD = 21.80). Grand total mean of Math GPA was 108.52 (SD = 26.26). Math GPA of high achievement sample was significantly higher than low achievement sample (F(1,361) = 448.47***). Within high achievement sample, males and females did not significantly differ from each other; within low achievement sample, males and females did not differ from each other either.

When Math GPA is calculated by gender, Math GPA of male sample was 112.41 (SD = 26.06); Math GPA of female sample was 104.47 (SD = 25.92). Math GPA of male sample was significantly higher than female sample (F (1,361) = 8.46**) in general.

- *Parental educational level of the whole sample and by achievement group*

SES levels are correlated with educational attainment, but it is the education that determines the kind of jobs held by people; thus it also determines their income and where they live. Because education appears to function as an independent variable that affects not only SES but a whole complex of values, attitudes, behavior patterns, many psychologists have come to prefer it as an index of social class. Hence in many research reports, SES should be more properly rendered as educational level. Psychologists have also found that the number of years of education completed is a more reliable measure than more complex scales that take into account occupational level, place of residence, and the other factors that are used as indicators of SES (Watson & Lindgren, 1979). Therefore, in the current study the researcher chose parental educational level as one of the major indicators for describing socioeconomic characteristics of the sample family.

Grand total mean of Mother's education level was 3.30 (SD = .91), which was between senior high school and Bachelor degree. Mother's education level of high achievement sample (M = 3.47, SD = .93) was significantly higher than that (M = 3.12, SD = .87) of low achievement sample (F (1,453) = 17.04***).

Grand total mean of Father's education level was 3.50 (SD = .89), which was between Senior high school and Bachelor degree. Father's education level of high achievement sample (M = 3.70, SD = .87) was significantly higher than that (M = 3.30, SD = .86) of low achievement sample (F (1,453) = 24.71***).

- *SES (Social economic status) of the whole sample and by achievement group*

Social economic status data were from participants' self-reports in the current study. Grand total mean of SES was 2.70 (SD = .81), which was between Not wealthy and Average families. SES mean of high achievement sample (M = 2.81, SD = .79) was significantly higher than that (M = 2.58, SD = .81) of low achievement sample (F (1,457) = 9.73**). This result was in line with the findings concerning parental educational level in the current research, which found that parental educational level of high achievers' were higher than that of low achievers.

The above results further confirmed the relation between SES and educational level (*Table 3.2.2*).

Among the whole sample, 459 participants revealed their SES data while 5 left it unanswered. The majority of the participants in the study were from average families (56.3%). The second biggest group was from not very wealthy families (21.6%). The third biggest group was from relatively wealthy families (11.0%). Participants from not at all wealthy families occupied the fourth order (9.9%), which was very close to the third biggest group. The smallest group of participants was from very wealthy families (0.2%). When the five SES groups were combined into three groups of not wealthy families, average families, and wealthy families, the valid percent of the three SES groups were 31.5%, 56.3%, 11.2% respectively.

Table 3.2.3: Social Economic Status Percentage of High and Low Achievement Samples

	Not at all wealthy	Not very wealthy	Average	Relatively wealthy	Very wealthy	Missing
Whole sample	9.9%	21.6%	56.3%	11%	0.2%	1.1%
High achievement sample	7.4%	19.5%	57.6%	14.7%	.4%	.4%
Low achievement sample	12.4%	23.6%	54.9%	7.3%	.0%	1.7%

The SES percentage of high achievement group equal or above average category was higher than low achievement group; The SES percentage of high achievement group equal or below the category of not very wealthy was lower than low achievement group (*Table 3.2.3*).

3.3 Measures

3.3.1 Choosing instruments

- *Rationale of choosing the measures*

The rationale of choosing the measures was closely connected with the purpose of the study, namely understanding adolescents' affective-cognitive processes in learning mathematics. It was more than a theoretical and empirical interest, but a part of sincere commitment to understand adolescents' psychological well-being and achievement, encompassing their achievement motivation, feelings and emotions, learning strategy use, etc. Therefore, instruments to measure adolescents' affective experience, learning strategy use, achievement goal orientation, values, math self-concept, and self-efficacy in math learning, etc., were chosen in the study. It is also known that students' motivation is influenced not only by their individual personal dispositions and beliefs, but also by the environment (Maehr, 1984; Nicholls, 1989). Ryan and Patrick (2001) found that the dimensions of teacher support, mutual respect, and promoting task-related interaction in the social context of the classroom may be an inextricable part of what it means to have a classroom emphasis on developing competence, or a mastery goal structure. Thus, instruments to measure students' immediate learning environment (parental control, peer attitude, teacher enthusiasm, perceived classroom climate) were chosen as part of the comprehensive measures, too.

- *Principles of choosing instruments*

The second issue concerning choosing instruments is how, namely under what principles were the instruments chosen for the study. Scott (1983) says there are no universally accepted answers to questions about the best method or approach to use with a particular vocational development problem. Whiteley (1978) notes that: "there is fundamental disagreement about which of the established methodologies, instruments, and counseling practices are most effective with clients (p.1)".

With the development and improvement of instruments, systematic study of individual differences became possible. Attention could be paid to social and ethical questions raised by the use of tests. We now ask not only "How able does the test show the candidate to be?" but also "How able was the candidate to show his ability on the test used?" and "How legitimate is it to make such an assessment in this case with this instrument?" (Super, 1971, p.7)

Two principles were considered in the present study during instrument choosing process: one is "culture differences" and the other is "conceptually reasonable and recognized instrument".

Principle 1: Culture difference Achievement emotions, motivation, learning strategy use, and expectation can be influenced by cultural factors. For example, separation and autonomy from the family are seen as mature and occupational choices are viewed as opportunities for self-expression and self-actualization in western societies, whereas eastern values center on collectivism and so decision-making occurs largely with one's community and family in mind (Gysbers, Heppner, & Johnston, 1998; Osipow & Fitzgerald, 1996). These cultural discrepancies can result in different processes and outcomes of achievement motivation, emotional experience in learning. Therefore, it is very imperative to examine in unbiased ways cultural differences in achievement motivation.

Adoption of culturally inappropriate instruments can cause problems, like misdiagnosis of problems (Fouad, 1993; Westermeyer, 1987). For measurement of applying learning strategies the Motivated Strategies for Learning (MSLQ; Pintrich et al., 1991) has been widely used today. AEQ-M (Academic Emotions Questionnaire for Mathematics) shows a high degree of measurement invariance across German and Chinese cultures (Frenzel et al., 2007c). AEQ-M was administered to Chinese samples (He, 2004; Ye, 2004; Wan, 2004) and was proved to be a reliable tool for measuring Chinese secondary school students' emotions in learning mathematics.

Principle 2: Conceptually reasonable and recognized instrument Pintrich and his colleagues' *Motivated Strategies for Learning Questionnaire* was chosen to collect students' learning strategy use data. The MSLQ was based on a 7-point Likert scale (1: not at all true of me 7: very true of me). MSLQ was developed through three stages of procedures (Pintrich et al., 1991). At stage one, a self-report instrument to assess students' motivation and use of learning strategies varying from 50 to 140 items was developed. At stage two, MSLQ was administered to obtain empirical data and consequently the conceptual model underlying the instrument was identified. At stage three, items and constructed scales were revised.

The *Academic Emotions Questionnaire – Mathematics* was chosen for the purpose of collecting students' affective experience data. Items for the AEQ-M were derived from the scales of the original Achievement Emotions Questionnaire (AEQ; Pekrun, Goetz, & Perry, 2005a; Pekrun et al., 2002a) by selecting items that are appropriate for students of grades 5 to 10, and are relevant for emotional experiences in mathematics (Pekrun, Goetz, & Frenzel, 2005b). Item analysis and scale revision were based on three consecutive studies using samples of German secondary school students (grades 5 to 10; Goetz, 2004). The final German AEQ-M scales were translated into the English language by a team of two bilingual experts. A back translation procedure was used to ensure content-related item equivalence (Pekrun et al., 2005b). The construct comparability of achievement emotions across German and Chinese students was established. In addition, the utility of the AEQ-M was established by examing the convergent validity. Regarding convergent validity, relationships between achievement emotions and several other achievement-related variables, including school achievement and attributions for success and failure (see Pekrun et al., 2002a) were examined. Similar patterns of relationships across German and Chinese samples supported the utility of the AEQ-M in cross-cultural research (Frenzel et al., 2007c).

The *Perceived Classroom Environment Questionnaire* was chosen for the purpose of collecting learning environment data. The learning environment in the study consisted of parental control, peer attitude and teacher enthusiasm three subscales besides perceived classroom environment. The 11 items for Perceived Classroom Environment Questionnaire were obtained through a pilot testing, which was conducted to create measures to assess the perceived classroom environment variables of central interest: lecture engagement, evaluation focus, and harsh evaluation (Church, Elliot, & Gable, 2001). The initial item pool consisted of revised items from existing classroom or sport climate scales (Ames & Archer, 1988; Frasier & Fisher, 1986; Winston, Vahala, Nichols, Gillis, Winthrow, & Rome, 1994), as well as new items generated for the present research (Church et al., 2001). Factor analysis validated the presence of the three perceived classroom environment factors (Church et al., 2001).

The *Achievement Goal Questionnaire* was chosen for the purpose of collecting students' goal orientation data. The 12 items for Achievement Goal Questionnaire (Elliot & McGregor, 2001) were systematically selected from the existing measures (Elliot, 1999; Elliot & Church, 1997) for mastery-approach, performance-approach, and performance-avoidance goals; new items were devised for mastery-avoidance goals. Pilot test was used to insure reliable and valid indexes of each of the four achievement goals in the 2 times 2 frameworks. Confirmatory factor analysis validated the presence of the four achievement goals (Elliot & McGregor, 2001).

Self-referenced cognitions are known as control beliefs and expectancy, referring to students' estimation of their ability (self-concept) and estimating the possibility of solving a learning problem (self-efficacy). Achievement values include intrinsic value and extrinsic value. In the study self-referenced cognitions and achievement values were centered on math learning scenario. The relevant measures were from Academic Emotions and Social Cognitions Questionnaire (AESCQ, Pekrun et al, 2002b).

3.3.2 Reliability of the measures

The major measures and their source, item numbers and reported reliability coefficients are in the table (*Table 3.3.1*).

Table 3.3.1: Source Information, Item numbers, and
Alpha Levels of Original Measures

Measure	Source	Items	Alpha
Motivated Strategies for Learning Questionnaire (Cognitive & Metacognitive self-regulation)	Pintrich et al., 1991	31	.64 to .80
Academic Emotions Questionnaire for Mathematics	Pekrun et al., 2005b	60	.84 to .92
Achievement Goal Questionnaire	Elliot & McGregor, 2001	12	.83 to .92
Perceived Classroom Environment Questionnaire	Church et al., 2001	11	.65 to .91
Causal Attribution Scale	Ames & Archer, 1988	12	
Self-referenced cognitions / achievement values	Pekrun et al., 2002b	8 / 10	.81 to .83 /.76 to .80

Note. The alphas of self-referenced cognitions and achievement values are from He's study (2004).

3.3.2.1 Motivated Strategies for Learning Questionnaire (MSLQ)

The 31 learning strategy items of MSLQ (Pintrich et al., 1991) includes cognitive and metacognitive learning strategies. The reported alpha levels of MSLQ (Cognitive and Metacognitive subscales) range from .64 to .80 (*Table 3.3.2*). Students rated themselves on a 7-point Likert scale from "not at all true of me" (1) to "very true of me" (7) on items concerning their use of cognitive and metacognitive learning strategies. Cognitive and Metacognitive learning strategies are made up of Rehearsal, Elaboration, Organization, Critical thinking, and Metacognitive self-regulation five subscales.

Table 3.3.2: Scale Descriptive Statistics, Internal Reliability Coefficients, and Correlations with Math GPA for all the Measures

Scale	M (SD)	Coefficient alpha	Reported alpha	r with Math GPA
Learning strategy				
Rehearsal	13.39 (4.55)	.55	.69	.07
Elaboration	23.44 (6.98)	.70	.76	.31
Organization	14.07 (4.80)	.67	.64	.19
Critical thinking	21.48 (6.77)	.78	.80	.41
Metacognitive self-regulation	52.95 (11.79)	.78	.79	.48
Academic emotions				
Enjoyment	32.10 (7.00)	.88	.90	.49
Pride	20.00 (4.53)	.82	.87	.37
Anger	19.67 (6.81)	.89	.88	-.49
Anxiety	38.12 (10.21)	.89	.92	-.41
Shame	20.39 (5.11)	.67	.84	-.20
Hopelessness	15.16 (5.08)	.85	.89	-.46
Boredom	13.67 (5.12)	.88	.89	-.55
Achievement goal				
Performance approach	14.57 (4.60)	.76	.92	.19
Mastery avoidance	15.36 (4.51)	.78	.89	-.01
Mastery approach	17.94 (3.37)	.74	.87	.31
Performance avoidance	12.42 (4.60)	.66	.83	-.29
Environmental factors				
Lecture engagement	18.11 (5.11)	.84	.91	.39
Harsh evaluation	16.48 (3.48)	.45	.74	-.29
Evaluation focus	11.32 (4.19)	.79	.65	-.24
Parental control	11.66 (4.86)	.87	.76	-.46
Peer attitude	9.41 (2.50)	.76	.76	.22
Teacher enthusiasm	14.11 (4.13)	.88	.75	.36
Self-referenced cognition				
Math self-concept	11.67 (2.80)	.76	.83	.39
Self-efficacy	13.05 (3.42)	.68	.81	.42
Achievement value				
Intrinsic value	18.50 (3.74)	.76	.80	.32
Extrinsic value	19.38 (3.81)	.77	.76	.11
Intrinsic parental value	9.59 (2.38)	.73	.78	.23
Extrinsic parental value	10.57 (2.42)	.69	.78	.10

Note. Source information for reported alphas is in Table 3.3.1. $|r| \geq .19$, $p < .001$; $|r| \geq .11$, $p < .05$.

Basic cognitive strategy – rehearsal Basic rehearsal strategies involve reciting or naming items from a list to be learned. These strategies are assumed to influence the attention and encoding processes, but they do not appear to help students construct internal connections among the information or integrate the information with prior knowledge (Pintrich et al., 1991). The Rehearsal subscale has 4 items (e.g., *"When studying for the math class, I read my class notes and the course readings over and over again."*).

Complex cognitive strategy – elaboration Elaboration strategies help students store information into long-term memory by building internal connections between items to be learned. Elaboration strategies include paraphrasing, summarizing, creating analogies, and generative note-taking. These help the learner integrate and connect new information with prior knowledge (Pintrich et al., 1991). The Elaboration subscale has 6 items (e.g., *"I try to relate ideas in the math subject to those in other courses whenever possible."*).

Complex cognitive strategy – organization Organization strategies help the learner select appropriate information and also construct connections among the information to be learned. Examples of organizing strategies are clustering, outlining, and selecting the main idea in reading passages. Organizing is an active, effortful endeavor, and results in the learner being closely involved in the task (Pintrich et al., 1991). The Organization subscale has 4 items (e.g., *"I make simple charts, diagrams, or tables to help me organize math course material."*).

Critical thinking Critical thinking refers to the degree to which students report applying previous knowledge to new situations in order to solve problems, reach decisions, or make critical evaluations with respect to standards of excellence (Pintrich et al., 1991). The Critical thinking subscale has 5 items (e.g., *"I often find myself questioning things I hear or read in the math course to decide if I find them convincing."*).

Metacognitive self-regulation Metacognition refers to the awareness, knowledge, and control of cognition. Control and self-regulation aspects of metacognition are focused on the MSLQ. There are three general processes making up metacognitive self-regulatory activities: planning, monitoring, and regulating. The Metacognitive self-regulation subscale has 12 items. Planning activities refer to goal setting which helps to activate relevant aspects of prior knowledge that make organizing and comprehending the material easier (e.g., *"When I study for this class, I set goals for myself in order to direct my activities in each study period."*). Monitoring activities include tracking of one's attention as one reads, and self-testing and questioning (e.g., *"I ask myself questions to make sure I understand the material I have been studying in this class."*). Regulating refers to the fine-tuning and continuous adjustment of one's cognitive activities (e.g., *"I try to change the way I study in order to fit the course requirements and instructor's teaching style."*). Regulating activities are assumed to improve performance by assisting learners in checking and correcting their behavior as they proceed on a task (Pintrich et al., 1991).

Reliability coefficients of Motivated Strategies for Learning Questionnaire obtained from the 464 sample in the study ranged from relatively low alpha level of .55 to moderate alpha level of .78 (*Table 3.3.2*). Among the five subscales, three subscale alphas were equal or above .70. The moderate alphas gave support that the Chinese version of the MSLQ was a reliable tool to measure students' applying learning strategies in learning math. The similar moderate levels of alphas of the Chinese version and the English version further revealed that the translation of the questionnaire from English to Chinese was relatively successful.

However, the reliability coefficient of the Rehearsal subscale of the Chinese version was only .55. The reliability coefficient of the Rehearsal subscale of the English version was .69, which was also not very high. The possible explanation of the lowered alpha in the Chinese version could be that when learning math students usually have to do exercises over and over again to really conquer math learning concepts. However, the Rehearsal items in the original English version was more suitable for checking literal arts subjects, e.g., language learning, and not very suitable for checking students' math learning. For example, item 41 *"I make lists of important terms for this course and memorize the lists."* The validity of the Rehearsal items to check students' math learning process was somehow problematic and therefore some more changes on the items should be done when a general questionnaire is used in a specific field.

3.3.2.2 Academic Emotions Questionnaire – Mathematics (AEQM)

The instrument measures emotions that are linked to achievement activities and their outcomes in the domain of mathematics, including enjoyment, pride, anger, anxiety, shame, hopelessness, and boredom. These emotions are experienced frequently by upper elementary, middle, and high school students in mathematics (Goetz, 2004).

AEQM has alpha levels of .84 to .92 (Pekrun et al., 2005b; **Table 3.3.1**). AEQM has seven sub-scales of Enjoyment, Pride, Anger, Anxiety, Shame, Hopelessness, and Boredom. Students rated themselves on a 5-point Likert scale from "strongly disagree" (1) to "strongly agree" (5) on items concerning their affective experience in class learning, doing homework, and sitting for a math exam situations.

Enjoyment The Enjoyment subscale has 10 items (e.g., *"When doing my math homework, I am in a good mood."*).

Pride The Pride subscale has 6 items (e.g., *"After a math test, I am proud of myself."*).

Anger The Anger subscale has 9 items (e.g., *"I am so angry that I would like to tear the exam paper into pieces."*).

Anxiety The Anxiety subscale has 15 items (e.g., *"When thinking about my mathematics class, I get nervous."*).

Shame The Shame subscale has 8 items (e.g., *"When I say something in my math class, I feel like embarrassing myself."*).

Hopelessness The Hopelessness subscale has 6 items (e.g., *"I keep thinking that I will never get good grades in mathematics."*).

Boredom The Boredom subscale has 6 items (e.g., *"Just thinking of my math homework assignments makes me feel bored."*).

Reliability coefficients of AEQM obtained based on the 464 sample data in the current study ranged from moderate alpha level of .67 to moderate high alpha level of .89 (***Table 3.3.2***). Among the seven subscales, six subscale alphas were equal or above .82. The satisfactory alphas gave strong support that the Chinese version of the AEQM was a very reliable tool to measure students' academic emotions in learning math. The similar high alpha levels of the

Chinese version and the English version further revealed that the translation of the questionnaire from English to Chinese was successful.

The reliability coefficient of the subscale Shame of the Chinese version was .67, which was much lower than the reliability coefficient of the Shame subscale of the English version .84. The alphas of the other six subscales were very similar across cultures. Cultural influence must have had some impact on the alpha differences of the Shame subscale.

Enjoyment ($r = .49$, $p < .001$) and pride ($r = .37$, $p < .001$) were positively related to math GPA, whereas anger ($r = -.49$, $p < .001$), anxiety ($r = -.41$, $p < .001$), shame ($r = -.20$, $p < .001$), hopelessness ($r = -.46$, $p < .001$), and boredom ($r = -.55$, $p < .001$) were negatively related to math GPA. These findings were consistent with previous research (e.g., Frenzel et al., 2007b).

3.3.2.3 Achievement Goal Questionnaire (AGQ)

Achievement Goal Questionnaire has alpha levels of .83 to .92 (Elliot & McGregor, 2001; **Table 3.3.1**). Achievement Goal Questionnaire has four sub-scales of Performance approach, Performance avoidance, Mastery approach, and Mastery avoidance. Students rated themselves on a 7-point Likert scale from "not at all true of me" (1) to "very true of me" (7) on items concerning their achievement goal orientation(s).

Mastery-approach goal In the mastery-approach goal construct, competence is defined in terms of the absolute requirements of the task or one's own pattern of attainment and competence is the focal point of regulatory attention (Elliot & Church, 1997; Elliot & Harackiewicz, 1996). The Mastery-approach goal subscale has four items (e.g., *"It is important for me to understand the content of this course as thoroughly as possible."*).

Mastery-avoidance goal In the mastery-avoidance goal construct, competence is defined in absolute/intrapersonal terms, and incompetence is the focal point of regulatory attention (Elliot & McGregor, 2001). The Mastery-avoidance goal subscale has four items (e.g., *"I worry that I may not learn all that I possibly could in this class."*).

Performance-approach goal Performance-approach goals are focused on the demonstration of competence relative to others, and competence is the focal point of regulatory attention (Elliot & Church, 1997; Elliot & Harackiewicz, 1996). The Performance-approach goal subscale has four items (e.g., *"It is important for me to do better than other students."*).

Performance-avoidance goal In the performance-avoidance goal construct, competence is defined in terms of normative terms, and incompetence is the focal point of regulatory attention (Elliot & Church, 1997; Elliot & Harackiewicz, 1996). The Performance-avoidance goal subscale has four items (e.g., *"I just want to avoid doing poorly in this class."*).

Reliability coefficients of Achievement Goal Questionnaire obtained from the 464 sample in the study ranged from moderate alpha level of .66 to moderate high alpha level of .78 (***Table 3.3.2***). Among the four subscales, three subscale alphas were equal or above .74. The moderate and moderate high alphas gave strong support that the Chinese version of the

AGQ was a very reliable tool to measure students' achievement goals in math learning process. However, the reported alphas of the original English questionnaire were higher compared with those of the Chinese version.

3.3.2.4 Perceived Classroom Environment Questionnaire (PCEQ)

The reported alpha levels for Perceived Classroom Environment Questionnaire ranged from .65 to .91 (Church et al., 2001). The perceived classroom environment subscales could be found within the category of environmental factors (**Table 3.3.2**). Perceived Classroom Environment Questionnaire has three sub-scales, namely Lecture engagement, Harsh evaluation, and Evaluation focus. Students rated themselves on a 7-point Likert scale from "strongly disagree" (1) to "strongly agree" (7) on items concerning their perceived classroom environment.

Lecture engagement Lecture engagement concerns the degree to which students perceive that the teacher makes the lecture material interesting (Church et al., 2001). Lectures that students find interesting and engaging are likely to facilitate "flow" and draw the student into the learning process (Brophy, 1986). Church and his colleagues (2001) believe that lecture engagement should promote the adoption of mastery goals; lecture engagement is expected to be unrelated to performance-approach and performance-avoidance goal adoption. The Lecture engagement subscale has 4 items (e.g., *"The teacher presents the material in an interesting manner."*).

Harsh evaluation Harsh evaluation concerns the extent to which students view the grading structure as so difficult that it minimizes the likelihood of successful performance. This form of evaluation is likely to result in performance-avoidance goals (Church et al., 2001). Harsh evaluation is also likely to evoke anxiety (Elliot, 1997), which is expected to reduce the likelihood of mastery goal adoption. Minimizing the availability of a positive outcome is likely to be negatively, or at minimum unrelated, to performance-approach goal adoption (Church et al., 2001). The Harsh evaluation subscale has 4 items (e.g., *"The grading structure makes it almost impossible to get an A in this course."*).

Evaluation focus Evaluation focus concerns the extent to which students perceive that the teacher emphasizes the importance of grades and performance evaluation in the course. A strong emphasis on evaluation is likely to orient students toward performance outcomes (Ames & Archer, 1988; Maehr & Midgley, 1991), which should prompt the adoption of both performance approach and avoidance goals (Church et al., 2001). The emphasis on external evaluation is also likely to discourage mastery pursuits (Ames, 1992; Meece, 1991). The Evaluation focus subscale has 3 items (e.g., *"The focus of the math course seems to be more on evaluating us than on teaching us."*).

Reliability coefficients of Perceived Classroom Environment Questionnaire obtained from the 464 sample in the study ranged from relatively low alpha level of .45 to moderate high alpha level of .84. Among the three subscales, two subscale alphas were equal or above .79. The moderate alphas gave support that the Chinese version of the PCEQ was a reliable tool to measure students' perception of classroom environment in math class. The similar alpha levels of the Chinese version and the English version further revealed that the translation of the questionnaire from English to Chinese was successful.

The other three environmental factors were measured using Social Cognitions Questionnaire (Pekrun et al., 2002b).

Parental control and punishment Parental control and punishment has 5 items (e.g., *"When I've got a bad grade in mathematics, my parents scold me. "*). Its reliability coefficient was .87.

Peer attitude Peer attitude has 3 items (e.g., *"Most of my classmates think that learning mathematics is worthwhile. "*). Its reliability coefficient was .76.

Teacher enthusiasm Teacher enthusiasm has 4 items (e.g., *"My maths teacher teaches with enthusiasm. "*). Its reliability coefficient was .88.

3.3.2.5 Self-Referenced Cognition and Achievement Values

Self-referenced cognition Self-referenced cognition scale is made up of math self-concept and self-efficacy in learning mathematics (Pekrun et al., 2002b). Students rated themselves on a 5-point Likert scale from "strongly disagree" (1) to "strongly agree" (5) on items concerning their math self-concept and self-efficacy beliefs. Previous study reported .83 (math self-concept) and .81 (self-efficacy in learning mathematics) by using this scale (He, 2004). Reliability coefficients of Self-referenced cognition obtained from the 464 sample in the study ranged from moderate alpha level of .68 to moderate high alpha level of .76 (***Table 3.3.2***). Self-referenced cognitions were positively correlated with math GPA. Math self-concept has 4 items (e.g., *"I am a talented student in mathematics. "*). Self-efficacy has 4 items (e.g., *"I am certain I can understand even most difficult contents of the course in mathematics. "*).

Achievement values Achievement value is constructed of intrinsic value and extrinsic value (Pekrun et al., 2002b). Students rated themselves on a 5-point Likert scale from "strongly disagree" (1) to "strongly agree" (5) on items concerning their intrinsic value and extrinsic value in learning mathematics. Intrinsic value has 5 items (e.g., *"No matter what grades I get, mathematics is very important to me. "*). Extrinsic value has 5 items (e.g., *"It is very important for me to obtain good grades in maths. "*). Previous study reported .80 (intrinsic value) and .76 (extrinsic value) by using this scale (He, 2004). Reliability coefficients of achievement value obtained from the 464 sample were moderate alpha levels of .76 and .77 (***Table 3.3.2***).

Intrinsic parental value has 3 items (e.g., *"In my family, we regard the subject of math as very important. "*). *Extrinsic parental value* has 3 items (e.g., *"In my family, it is important to be good at mathematics. "*). The reliability coefficient alpha of intrinsic parental value was .73. The reliability coefficient alpha of extrinsic parental value was .69 (***Table 3.3.2***).

The moderate alphas gave strong support that the Chinese versions of the Self-Referenced Cognition and Achievement Values were reliable tools to measure students' self-referenced cognition and achievement values in math.

3.3.2.6 Causal Attribution Questionnaire

Causal attribution scale was from Ames and Archer (1988). Students were asked two sets of attribution questions related to when they did well and not very well in class. For each set, students rated the importance of ability (have ability, not have enough ability), effort (worked very hard, did not work hard enough), strategy (used good strategies, did not use good strategies), the task (work was easy, work was difficult), and the teacher (teacher did a good job, teacher did a poor job) as reasons for their performance. Five-point scales (1=not an important reason; 5= an important reason) were used for each rating. The researcher added luck (had good luck, did not have good luck) as another dimension of causal attribution.

Since for each attribution there was only one statement to be measured, coefficient alphas were not possible to be calculated. However, means for each attribution provided some chance looking into the causal attribution patterns of the participants.

3.3.3 Predictive validity of the measures

The predictive validity analyses included correlations of the major scales with math academic performance in terms of Math GPA (grand point average). The scale descriptive statistics were described in *Table 3.3.2*. The scales showed significant correlations with Math GPA (The sample size was 360; Correlations of .19 and above were significant at .001 level and correlations of .11 and above were significant at .05 level.). The correlations were in the expected directions, adding to the validity of the scales. Students who had more positive affect, applied more learning strategies except for Rehearsal, took approach goals, reported more control of their learning, and valued math learning were more likely to get higher Math GPA. At the same time, students who reported experiencing more negative affect, and took performance avoidance goal were less likely to achieve high Math GPA. It was also found out that positive environmental factors were positively correlated with Math GPA, whereas negative environmental factors were negatively correlated with Math GPA. Among the environmental factors, lecture engagement was positively correlated with Math GPA ($r = .39$, $p < .001$), whereas harsh evaluation ($r = -.29$, $p < .001$) and evaluation focus ($r = -.24$, $p < .001$) were negatively correlated with Math GPA. In contrast, Church et al. (2001) found that the three dimensions were not related to students' graded performance.

3.3.4 Translation of the questionnaires

According to Butcher's (1982) steps for preparing a translation, the following steps (translation, back translation, and field testing) were taken to insure a valid formation of Chinese versions of the applied questionnaires. Two bilingual translators, the researcher herself and a PhD student whose native language is Chinese, worked independently and each prepared two initial translations of the inventories. The two versions of each inventory were reviewed by the other translator. After some terms were corrected and modified, initial translations of the two inventories were created. A back translation procedure to detect any difficulties or failed communication was carried out. Back translation also operates as a filter in which non-equivalent terms are screened out (Schrest, Fay, & Zaidi, 1972). A third PhD bilingual student translated the Chinese versions of the above questionnaires back into English versions. Concrete information concerning field testing was presented in the following subchapters of translation of the questionnaires.

Translation of the MSLQ (Motivated Strategies for Learning Questionnaire)

There were minor changes in several items in the Chinese version of the MSLQ to make the questionnaire more suitable for measuring the field of mathematics learning. Example 1: "When I study the readings for this course, I outline the material to help me organize my thoughts" was changed to "When I study the content for this course, I outline the material to help me organize my thoughts". These changes were made so that the questionnaire would better reflect (a) the practice of mathematics learning and (b) the classroom environment in China.

- *Field testing*

After translation, back translation, the last step taken to check the validity of the MSLQ translation was a bilingual field testing which examined correspondence between the English version and the translated Chinese version. Sixty students (one intact class) of the elite senior high school sample participated in this bilingual testing. However, only 23 students' data was finally processed to check the consistency of the two versions of the questionnaire. Those two versions were administered 10 days apart. They rated themselves for each item on a 7-point Likert scale (1 和我的情形完全不符: 1 not at all true of me to 7 和我的情形十分相似: 7 very true of me).

Table 3.3.3: Descriptive Statistics (Means and Standard Deviations), t Values for the English and the Chinese Version of MSLQ

	English version	Chinese version	T
Rehearsal	4.39(1.27)	3.33(.94)	5.31***
Elaboration	4.71(.84)	3.93(.91)	3.66***
Organization	4.30(1.53)	3.60(1.29)	3.17**
Critical thinking	4.93(.86)	4.60(1.12)	1.88
Metacognitive self-regulation	5.28(.72)	4.78(.76)	4.58***

Note: The unit of analysis is judge (N = 23). Standard deviation is in parentheses.
**p < .01 ***p < .001
MSLQ: Motivated Strategies for Learning Questionnaire (Pintrich et al., 1991)

- *t-test*

Differences between the two versions for each of the 5 subfactors were statistically significant except for critical thinking (*Table 3.3.3*). Mean scores for each subfactors of the English version were higher than mean scores of the Chinese version. The participants seemed to have overestimated their use of learning strategies when they were confronted with the English questionnaire. One possible explanation for the significant higher scores could be that the order effect on the participants' responses was not controlled over through counter-balancing. Consequently, participants got familiar with the questionnaire items through filling out the Chinese version before they were administered the English version ten days later.

- *Correlations between the two versions*

Table 3.3.4: A Correlation Matrix for Five Constructs of the English and the Chinese Version of MSLQ

Variables	1	2	3	4	5	6	7	8	9
English version									
1 Rehearsal									
2 Elaboration	.79**								
3 Organization	.78**	.61**							
4 Critical thinking	.54**	.70**	.27						
5 Metacognitive self-regulation	.68**	.67**	.53**	.43*					
Chinese version									
6 Rehearsal	.66**	.42*	.66**	.23	.37				
7 Elaboration	.41	.32	.69**	.02	.34	.46*			
8 Organization	.53*	.42*	.73**	-.01	.33	.66**	.58**		
9 Critical thinking	.37	.46*	.15	.67**	.42*	.26	-.01	.03	
10 Metacognitive self-regulation	.54*	.42*	.48*	.28	.75**	.33	.31	.28	.23

*Note: The unit of analysis is judge (N = 23). *p < .05 **p < .01*
MSLQ: Motivated Strategies for Learning Questionnaire (Pintrich et al., 1991)

To investigate similarity between and validity of the two versions of MSLQ, bivariate correlations were done (*Table 3.3.4*). Although results from t-tests showed higher scores for sub-factors based on the English version, correlation results were supportive and the data revealed that correlations between those two versions for each sub-factor ranged from .66 to .75 except for elaboration.

Translation of the AEQM (Academic Emotions Questionnaire for Mathematics)

- *Field testing*

The last step taken to check the validity of the AEQM translation was a bilingual field testing which examined correspondence between the English version and the Chinese-translated version. Sixty students (one intact class) of the elite senior high school sample participated in this bilingual testing. However, only 23 students' data was finally processed to check the consistency of the two versions of the questionnaire. Those two versions were administered 10 days apart. Responses to items were measured on a 5-point Likert scale (1 十分不赞同: 1 strongly disagree to 5 十分赞同: 5 strongly agree).

Table 3.3.5: Descriptive Statistics (Means and Standard Deviations) and t Values for the English and the Chinese Version of AEQM

	English version	Chinese version	T
Enjoyment	3.68(.59)	3.53(.43)	1.97
Pride	3.60(.65)	3.58(.69)	.15
Anger	1.70(.58)	1.73(.58)	-.36
Anxiety	2.32(.68)	2.39(.56)	-.80
Shame	2.53(.56)	2.46(.52)	.73
Hopelessness	1.88(.83)	2.26(.76)	-2.67*
Boredom	1.59(.68)	1.70(.67)	-1.46

Note: The unit of analysis is judge (N = 23). Standard deviation is in parentheses.
*p < .05
AEQM: Academic Emotions Questionnaire for Mathematics (Pekrun, et. al., 2005b)

Table 3.3.6: A Correlation Matrix for Seven Constructs of the English and the Chinese Version of AEQM

Variables	1	2	3	4	5	6	7	8	9	10	11	12	13
English version													
1 Enjoyment													
2 Pride	.65**												
3 Anger	-.64**	-.79**											
4 Anxiety	-.59**	-.50**	.62**										
5 Shame	-.40	-.55**	.62**	.55**									
6 Hopelessness	-.55**	-.49**	.59**	.87**	.64**								
7 Boredom	-.59**	-.37	.55**	.57**	.25	.53**							
Chinese version													
8 Enjoyment	.81**	.50*	-.52*	-.57**	-.19	-.38	-.64**						
9 Pride	.36	.46*	-.30	-.24	-.19	-.21	-.33	.38					
10 Anger	-.44*	-.42*	.60**	.38	.26	.36	.66**	-.52*	-.58**				
11 Anxiety	-.45*	-.42*	.53**	.79**	.36	.66**	.57**	-.51*	-.34	.62**			
12 Shame	-.41	-.54**	.59**	.56**	.63**	.65**	.43**	-.32	-.14	.53**	.62**		
13 Hopelessness	-.35	-.35	.45**	.71**	.24	.64**	.50*	-.47	-.45	.63**	.84**	.44*	
14 Boredom	-.60**	-.33	.43*	.57**	.10	.44*	.86**	-.72**	-.28	.69**	.63**	.50*	.52*

Note: The unit of analysis is judge (N = 23). *p < .05 **p < .01.
AEQM: Academic Emotions Questionnaire for Mathematics (Pekrun, et. al., 2005b)

- *t-test*

Differences between the two versions for each of the 7 subfactors were not statistically significant except for hopelessness at .05 level *(Table 3.3.5)*.

- *Correlations between the two versions*

To investigate similarity between and validity of the two versions of AEQM, bivariate correlations were done *(Table 3.3.6)*. The data revealed that correlations between those two versions for each sub-factor ranged from .46 to .86. All submatrices of intertrait correlations had a similar pattern.

Translation of the AGQ (Achievement Goal Questionnaire)

- *Field testing*

The last step taken to check the validity of the AGQ translation was a bilingual field testing which examined correspondence between the English and the Chinese versions. Sixty students (one intact class) of the elite senior high school sample participated in this bilingual testing. However, only 23 students' data was finally processed to check the consistency of the two versions. Those two versions were administered 10 days apart. They rated themselves for each item on a 7-point Likert scale (1 和我的情形完全不符: 1 not at all true of me to 7 和我的情形十分相似: 7 very true of me).

Table 3.3.7: Descriptive Statistics (Means and Standard Deviations), t Values for the English and the Chinese Version of AGQ

	English version	Chinese version	T
Performance approach	5.88(.72)	5.51(1.27)	2.12*
Mastery avoidance	5.16(1.33)	5.35(1.29)	-.82
Mastery approach	6.42(.70)	6.54(.56)	-.79
Performance avoidance	3.35(1.36)	3.67(1.37)	-1.09

Note: The unit of analysis is judge (N = 23). Standard deviation is in parentheses.
*p < .05. AGQ: Achievement Goal Questionnaire (Elliot, 2001)

- *t-test*

Differences between the two versions for each of the 4 subfactors were not significant except for performance approach goal (Mean $_{English}$ = 5.88(.72), Mean $_{Chinese}$ = 5.51(1.27), t = 2.12*) *(Table 3.3.7)*.

- *Correlations between the two versions*

To investigate similarity between and validity of the two versions of AGQ, bivariate correlations were done *(Table 3.3.8)*. The data revealed that correlations between those two versions for each sub-factor ranged from .47 to .77 except for nonsignificant correlation between mastery approaches. Also all submatrices of intertrait correlations had a similar pattern.

Table 3.3.8: A Correlation Matrix for Four Constructs of the English and the Chinese Version of AGQ

Variables	1	2	3	4	5	6	7
English version							
1 Performance approach							
2 Mastery avoidance	.42*						
3 Mastery approach	.44*	.46*					
4 Performance avoidance	.37	.18	.08				
Chinese version							
5 Performance approach	.77**	.54**	.43*	.18			
6 Mastery avoidance	.53**	.65**	.51*	.24	.62**		
7 Mastery approach	.36	.51*	.39	-.05	.58**	.61**	
8 Performance avoidance	.53**	.55**	.32	.47*	53**	.59**	.21

Note: The unit of analysis is judge (N = 23).
$p < .05$ **$p < .01$(2-tailed). AGQ: Achievement Goal Questionnaire (Elliot, 2001)

Translation of the PCEQ (Perceived Classroom Environment Questionnaire)

- *Field testing*

The last step taken to check the validity of the PCEQ translation was a bilingual field testing which examined correspondence between the English version and the Chinese-translated version. Sixty students (one intact class) of the elite senior high school sample participated in this bilingual testing. However, only 23 students' data was finally processed to check the consistency of the two versions of the questionnaire. Those two versions were administered 10 days apart. They rated themselves for each item on a 7-point Likert scale (1 十分不赞同: 1 strongly disagree to 7 十分赞同: 7 strongly agree).

- *t-test*

Table 3.3.9: Descriptive Statistics (Means and Standard Deviations), t Values for the English and the Chinese Version of PCEQ

	English version	Chinese version	T
Lecture engagement	5.36(1.06)	4.93(.43)	1.85
Harsh evaluation	3.49(.83)	3.72(.77)	-1.00
Evaluation focus	4.10(1.38)	3.52(1.06)	2.01

Note: The unit of analysis is judge (N = 23). Standard deviation is in parentheses. PCEQ: Perceived Classroom Environment Questionnaire (Church, Elliot, and Gable, 2001)

Differences between the two versions for each of the 3 subfactors were not statistically significant *(Table 3.3.9)*.

- *Correlations between the two versions*

Table 3.3.10: A Correlation Matrix for Three Constructs of the English and the Chinese Version of PCEQ

Variables	1	2	3	4	5
English version					
1 Lecture engagement					
2 Harsh evaluation	-.50*				
3 Evaluation focus	-.18	.42*			
Chinese version					
4 Lecture engagement	.44*	-.004	.22		
5 Harsh evaluation	-.22	.06	.18	.25	
6 Evaluation focus	-.38	.21	.38	-.37	.30

Note: The unit of analysis is judge (N = 23). *p < .05
PCEQ: Perceived Classroom Environment Questionnaire (Church, Elliot, and Gable, 2001)

To investigate similarity between and validity of the two versions of PCEQ, bivariate correlations were done *(Table 3.3.10)*. The data revealed that for each sub-factor only lecture engagement was significantly correlated (r = .44 at .05 level).

3.4 Procedures

3.4.1 Sampling procedure

The data for the cross-sectional study was collected from an elite/key senior high school and two ordinary senior high schools during the end of March and the middle of April, 2007 in two metropolitans, Shi Jiazhuang of Hebei province (north China) and Wuhan of Hubei province (south China).

Half of the data, namely high achievement sample (N = 231, M = 17.36, SD = .52) was collected from the key senior high school. Participants of four intact classes of Grade two (similar to Grade 11 in the English school system) were first given the translated Chinese version of the questionnaire. Ten days later, sixty participants (one intact class) of the high achievement sample were given the English version of the questionnaire. However, only 23 of the 60 participants' filled-out English questionnaires were used to check the consistency of the two versions of the questionnaire and reliability of the Chinese version. The cut-off criteria was that participants' English achievement ranged equal or above A (135 out of 150 scores). A flaw in the process of collecting data for bilingual testing was that the order of administering the translated and original questionnaires was not controlled for through counter-balancing strategy. Consequently, participants got familiar with the questionnaire items before they filled out the English version.

Half of the data, namely low achievement sample ($N = 233$, $M = 16.96$, $SD = .53$) was collected from the two ordinary senior high schools. Participants of four intact classes of Grade two (similar to Grade 11 in the English school system) from the two ordinary high schools were only administered the Chinese version of the questionnaire.

3.4.2 Technical details concerning data collection procedure

Both the Chinese version questionnaire and the English version questionnaire were answered in a 45 minute class session. After the questionnaires were handed out to the participants, introductions and instructions were given by the researcher. The introductions included the purpose of the study, namely analysizing adolescents' affective and cognitive processes in learning mathematics, etc. The participants were informed that there were no correct or wrong answers and different choices only reflected individual differences. No discussions with classmates were encouraged. However, questions to the researcher were welcome. During participants answering the English version questionnaire, questions concerning meanings of new vocabulary were also encouraged. Participants were also informed that individual feedback was possible if participants left their email address at the end of the questionnaire.

3.4.3 Observation of the data collection process

The majority of the high achievement sample (231 participants) finished answering the Chinese version in 20 to 30 minutes. The questionnaires were collected after the class was over. Within the high achievement sample 60 students (one intact class) took part in the English questionnaire survey ten days later. The majority of the 60 participants finished answering the English version in 35 to 45 minutes. Six participants were not able to finish answering the English questionnaire within 45 minutes therefore 10 extra minutes were given to them. The majority of the moderate/low achievement sample (233 participants) finished answering the Chinese questionnaire in 30 to 40 minutes. The questionnaires were collected after the class was over.

Chapter IV   RESULTS

4.1   Validity of the measures

Statistical results concerning validity of the measures were not supposed to answer the research questions or hypotheses but to validate the measures applied in a collectivist culture of China in the current study.

4.1.1 Factor structure of MSLQ

Factor analytic techniques were conducted to test whether the dimensions of learning strategies could be validated with the Chinese sample. The data showed substantial common variance (KMO = .91) and correlations (Bartlett-test of sphericity, Chi-square = 4560.70, df = 465, p < .001). An EFA (exploratory factor analysis) was conducted on the 31 learning strategy items using principal axis factoring with promax rotation.

The analysis yielded six factors with an Eigen value exceeding 1, and the factor solution accounted for 51.77% of the total variance. Factor 1 accounted for 27.29% of the variance and consisted of the critical thinking and metacognitive items plus 2 cognitive items (1 rehearsal and 1 elaboration strategy) (Eigen value = 8.46). Factor 2 accounted for 7.46% of the variance and comprised 6 cognitive items (4 items were complex cognitive strategies) plus 1 metacognitive item (Eigen value = 2.31). The third factor accounted for 5.79% of the variance and comprised 3 cognitive (2 items were complex cognitive strategies) and 2 metacognitive items (Eigen value = 1.79). Factor 4 accounted for 3.95% of the variance and comprised 2 metacognitive and 3 cognitive items (3 elaboration items) (Eigen value = 1.22). Factor 5 accounted for 3.92% of the variance and comprised 2 metacognitive items (Eigen value = 1.22). Factor 6 accounted for 3.37% of the variance and consisted of 1 metacognitive item (Eigen value = 1.05).

Critical thinking and some of the metacognitive items were loaded as one factor. Cognitive and metacognitive coexisted in factor 2, 3, and 4. Factor 5 and 6 were composed of all together 3 metacognitve items. The above factor structure of the learning strategies obtained from the Chinese sample gave support that critical thinking and metacognitive were close to each other although some metacognitive strategies were loaded on cognitive strategy scales. The mixture of metacognitive and cognitive especially cognitive complex strategies further reflected that the functions of some complex cognitive strategies were somehow approaching that of metacognitive strategies.

4.1.2 Factor structure of AEQM

Factor analytic techniques were conducted to test whether the seven dimensions of academic emotions could be validated with the Chinese sample. The data showed substantial common variance (KMO = .95) and correlations (Bartlett-test of sphericity, Chi-square = 16394.98, df = 1770, p < .001). An EFA was conducted on the 60 academic emotion items using maximum likelihood with promax rotation.

The analysis yielded seven factors with an Eigen value exceeding 1, and the factor solution accounted for 56.02% of the total variance. The chi-square to degrees of freedom ratio (χ^2/df) was 1.85. A χ^2/df ratio of less than 5 is considered to be indicative of a good fit between the observed and reproduced correlation matrices (Hayduk, 1987). Factor 1 accounted for 33.02% of the variance and consisted of mainly the anger and boredom items plus 4 anxiety items and 2 shame items (Eigen value = 19.81). Factor 2 accounted for 8.68% of the variance and comprised the enjoyment and pride items plus 1 anxiety item (Eigen value = 5.21). The third factor accounted for 4.24% of the variance and comprised 5 shame items, 4 anxiety items, and 2 hopeless items (Eigen value = 2.54). Factor 4 accounted for 2.93% of the variance and comprised 2 anxiety and 1 shame item (Eigen value = 1.76). Factor 5 accounted for 2.68% of the variance and comprised 4 anxiety and 3 hopelessness items (Eigen value = 1.61). Factor 6 accounted for 2.38% of the variance and consisted of 1 enjoyment and 1 pride item (Eigen value = 1.43). Factor 7 accounted 2.10% of the variance and comprised 1 hopelessness and 1 boredom item (Eigen value = 1.26).

Shame, hopelessness, and anxiety items were found to coexist in Factor 3, Factor 4, Factor 5, which further confirmed the reality that the three emotions were inseparable or closely connected with one another in math learning context in China. The three emotions were inwardly directed, reflecting limited resources to cope with outward demand. Ability was predicted to be the most salient predictor for shame, hopelessness, and anxiety. Anger and boredom were inherently different from shame, hopelessness, and anxiety. Anger and boredom were emotions mainly directing outward. Teaching quality and task difficulty were very likely to be responsible for anger and boredom. Enjoyment and pride, the two positive emotions, were loaded on one factor and were separated from the rest negative emotions.

Similar factor structures were found in He's study (2004) with a Chinese sample self reporting their emotions in learning mathematics. Anger and boredom were loaded on Factor 1, shame, hopelessness, and anxiety were loaded on factor 2, and enjoyment and pride were loaded on factor 3.

The factor structure of the Chinese sample's academic emotions confirmed the existence of positive and negative emotions in learning mathematics and differentiated the inwardly directing negative emotions from the outwardly directing negative emotions. Therefore, the AEQM was a valid tool to measure Chinese students' emotions in learning mathematics.

4.1.3 Factor structure of AGQ

Factor analytic techniques were conducted to test whether the four dimensions of achievement goals could be validated with the Chinese sample. It was predicted that exploratory factor analysis (EFA) of the achievement goal items would yield four distinct factors corresponding to the four dimensions.

Table 4.1.1: Factor Loadings for Achievement Goals

Achievement goal item	Factor		
	Mastery goal	Performance avoidance	Performance approach
1.I want to learn as much as possible from the math class.	.76		
2.It is important for me to understand the content of the math class as thoroughly as possible.	.73		
3.I desire to completely master the material presented in the math class.	.66		
4.I worry that I may not learn all that I possibly could in the math class.	.78		
5.I am often concerned that I may not learn all that there is to learn in the math class.	.67		
6.Sometimes I'm afraid that I may not understand the content of the math class as thoroughly as I'd like.	.68		
7.I just want to avoid doing poorly in the math class.		.78	
8.My fear of performing poorly in the math class is often what motivates me.		.69	
9.My goal in the math class is to avoid performing poorly.		.75	
10.It is important for me to do better than other students.			.68
11.It is important for me to do well compared to others in the math class.			.85
12.My goal in the math class is to get a better grade than most of the other students.			.77

Note. N = 460. All factor loadings > .33 are presented in the table. Factor loadings were obtained using principal component extraction with oblique rotation.

An EFA was conducted on the 12 perceived classroom environment items using principal component extraction with oblique (promax) rotation. The data showed substantial common variance (KMO = .84) and correlations (Bartlett-test of sphericity, Chi-square = 1874.64, df = 66, p < .001).

The analysis yielded three factors with an Eigen value exceeding 1, and the factor solution accounted for 61.42% of the total variance. *Table 4.1.1* displays the loadings for each factor. Factor 1 accounted for 36.59% of the variance and consisted of the three mastery approach and three mastery avoidance items (Eigen value = 4.39). Factor 2 accounted for 14.61% of the variance and comprised the three performance avoidance items (Eigen value = 1.75). The third factor accounted for 10.22% of the variance and comprised the three performance approach items (Eigen value = 1.23). None of the secondary loadings exceeded .32. The three factors represent separable achievement goal constructs with mastery approach and mastery avoidance items being loaded on one factor of mastery goal and performance approach and performance avoidance items being loaded on as another two distinct factors. Thus, the factors of achievement goals found with the Chinese sample were in line with the factor construct of the Elliot and Church's (1997) hierarchical model of approach and avoidance achievement motivation.

Table 4.1.2: Descriptive Statistics, Reliabilities, and Intercorrelations among Achievement Goal Variables

Variable	M	SD	Observed range	Possible range	Cronbach's α	Variable 1	2	3
1. Mastery goal	5.55	1.16	1-7	1-7	.82	–		
2. Performance avoidance	4.14	1.53	1-7	1-7	.66	.33**	–	
3 Performance approach	4.86	1.53	1-7	1-7	.76	.50**	.30**	–

Note. **$p < .01$ (2-tailed).

Moderate to moderate high reliabilities were found for performance approach and mastery goals. Performance avoidance goal had a relatively low reliability coefficient of .66. Mastery goal was positively associated with performance avoidance goal ($r = .33$, $p < .01$) and positively associated with performance approach goal ($r = .50$, $p < .01$). Performance avoidance goal was positively associated with performance approach goal ($r = .30$, $p < .01$; *Table 4.1.2*). In the previous study, mastery approach goal was not correlated with performance avoidance goal significantly (Elliot & McGregor, 2001).

4.1.4 Factor structure of PCEQ

Factor analytic techniques were conducted to test whether the three dimensions of perceived classroom environment could be validated with the Chinese sample. It was predicted that exploratory factor analysis (EFA) of the perceived classroom environment items would yield three distinct factors corresponding to the three dimensions.

An EFA was conducted on the 11 perceived classroom environment items using principal-axis factoring with promax rotation. The data showed substantial common variance (KMO = .81) and correlations (Bartlett-test of sphericity, Chi-square = 1901.35, df = 55, p < .001).

The analysis yielded three factors with an Eigen value exceeding 1, and the factor solution accounted for 63.11% of the total variance. *Table 4.1.3* displays the loadings for each factor. Factor 1 accounted for 38.26% of the variance and consisted of the three evaluation focus items plus the two harsh evaluation items absent in the third factor (Eigen value = 4.21). Factor 2 accounted for 14.69% of the variance and comprised the four lecture engagement items (Eigen value = 1.62). The third factor accounted for 10.15% of the variance and comprised the two of the four harsh evaluation items (Eigen value = 1.12). None of the secondary loadings exceeded .30. The three factors represent separable perceived classroom environment constructs although two of harsh evaluation items were loaded on evaluation focus subscale. Thus, the factors of perceived classroom environment found with the Chinese sample were congruent with the factor construct of the original measure.

Table 4.1.3: Factor Loadings for Perceived Classroom Environment

	Factor		
Perceived Classroom Environment item	Evaluation focus	Lecture engagement	Harsh evaluation
1.The focus of this course seems to be more on evaluating us than on teaching us.	.59		
2.The grading structure makes it almost impossible to get an A in this course.	.44		
3.The professor is more concerned with our grades than what we learn.	.81		
4.High grades are seldom obtained by students in this course.	.53		
5.It seems like all the professor cares about is how we do on the exams.	.86		
6.I find the lectures to be very engaging.		.74	
7.The way that the professor helps us learn the material holds my interest.		.88	
8.The professor presents the material in an interesting manner.		.70	
9.I find the lectures boring.		.57	
10.The grading for this course is pretty harsh.			.79
11.The grading in this class is pretty easy.			.46

Note. N = 462. All factor loadings > .30 are presented in the table. Factor loadings were obtained using principal axis factoring with oblique rotation.

Moderate to moderate high reliabilities were found for evaluation focus and lecture engagement. Harsh evaluation had a relatively low reliability coefficient of .51. Evaluation focus was negatively associated with lecture engagement (r = -.54, p < .01) and positively associated with harsh evaluation (r = .10, p < .05). Harsh evaluation was not associated with lecture engagement significantly (*Table 4.1.4*). In the previous study, harsh evaluation was found to be negatively correlated with lecture engagement significantly (Church et al., 2001).

Table 4.1.4: Descriptive Statistics, Reliabilities, and Intercorrelations among Perceived Classroom Environment Variables

						Variable		
Variable	M	SD	Observed range	Possible range	Cronbach's α	*1*	*2*	*3*
1. Evaluation focus	3.52	1.18	1-7	1-7	.78	_		
2. Lecture engagement	4.53	1.28	1-7	1-7	.84	-.54**	_	
3. Harsh evaluation	5.10	1.03	1-7	1-7	.51	.10*	.08	_

Note. $*p < .05$, $**p < .01$ *(2-tailed).*

4.2 Multivariate analyses of variance and covariance

All the major variables were split into two groups to be analysized by multivariate analyses of variance, namely Group 1 including attitude toward mathematics, math self-concept, self-efficacy in learning mathematics, intrinsic value of learning mathematics, extrinsic value of learning mathematics, self achievement expectation, and achievement expectation from parents, and Group 2 including 7 academic emotions, 5 learning strategies, and 4 achievement goals. Group 1 variables did not have sub-level variables, whereas Group 2 variables had sub-level variables.

First multivariate analyses of variance (MANOVA) and covariance (MANCOVA) were applied to Group 1 and Group 2 variables by achievement and gender respectively, since the researcher was interested to find out whether the means of the major variables were significantly different by achievement and gender respectively. In the case of MANOVA, ability group was used as independent variable, and Group 1 and Group 2 as dependent variables. In the case of MANCOVA, prior mathematics achievement (mathematics grade) was used as covariate, gender as independent variable, and Group 1 and Group 2 as dependent variables.

Then two multivariate analyses of variance were conducted to check achievement by gender effect on the Group 1 and Group 2 variables. The MANOVAs were performed using the

Group 1 and Group 2 variables as dependent variables and ability groups (high achievement group vs. moderate and low achievement group) and gender as independent variables.

When interaction effects were found significant, further analyses were conducted to compute simple effects and simple contrasts for comparing group differences within each level of an independent variable. When main effects were significant, necessary post hoc multiple comparisons were performed using the Tukey HSD test.

Assumptions for MANOVA procedures were carefully examined to determine if they were reasonable analytical procedures for the current data. Skewness measured for each variable for each cell was not extreme, mostly smaller than the absolute value of 1 with the largest value of -1.782. Scatter plots for each variable within each group revealed no cause for concern for linearity.

Box's M test of equality of covariance matrices was significant for the Group 1 variables. Univariate tests of homogeneity of variance partially indicated equality of variances for individual dependent variables except for attitude toward math, intrinsic parental value, self achievement expectation, and achievement expectation from parents, $p < .05$. For the Group 2 variables, Box's M test was significant with enjoyment, anger, boredom, critical thinking, metacognitive learning strategy, and mastery approach indicating nonequality of variance, $p < .05$. Thus, Pillai's criterion was used to evaluate multivariate significance (Mertler & Vannatta, 2001). The overall evaluations of assumptions were considered acceptable to continue with the following statistical procedures.

4.2.1 Attitude toward math, self, achievement values, and expectations

MANOVA of the variables by achievement

- *Attitude toward math by achievement*

Attitude toward math of high achievement sample ($M = 3.74$, $SD = .81$) was significantly higher than that of ($M = 3.23$, $SD = .96$) low achievement sample ($F (1, 458) = 38.72$, $p < .001$).

- *Math self-concept and self-efficacy by achievement*

Math self-concept of high achievement sample ($M = 3.09$, $SD = .64$) was significantly higher than that ($M = 2.74$, $SD = .71$) of low achievement sample ($F (1, 461) = 31.67$, $p < .001$). Math self-efficacy of high achievement sample ($M = 3.46$, $SD = .72$) was significantly higher than that ($M = 3.04$, $SD = .79$) of low achievement sample ($F (1, 461) = 35.56$, $p < .001$).

- *Intrinsic value and extrinsic value in math learning by achievement*

Intrinsic value of high achievement sample ($M = 3.87$, $SD = .69$) was significantly higher than that ($M = 3.54$, $SD = .77$) of low achievement sample ($F (1, 461) = 23.69$, $p < .001$). Extrinsic value of high achievement sample ($M = 3.95$, $SD = .77$) was significantly higher than that ($M = 3.81$, $SD = .75$) of low achievement sample ($F (1, 461) = 3.86$, $p < .01$).

- *Intrinsic and extrinsic parental value of learning math by achievement*

Intrinsic parental value of high achievement sample (M = 3.33, SD = .81) was significantly higher than that (M = 3.06, SD =. 75) of low achievement sample (F (1, 456) = 12.97, p < .001). Extrinsic parental value of high achievement sample (M = 3.60, SD = .82) was significantly higher than that of (M = 3.45, SD = .79) low achievement sample (F (1, 456) = 4.20, p < .01).

Table 4.2.1: Means and Standard Deviations of Attitude toward Math, Self, Values, and Achievement Expectations of High and Low Achievement Samples

	High achievement group		Low achievement group		Grand total			F
	Mean	*SD*	*Mean*	*SD*	*Mean*	*N*	*SD*	
Attitude toward math	3.74	.81	3.23	.96	3.48	460	.93	38.72***
Math self-concept	3.09	.64	2.74	.71	2.92	463	.70	31.67***
Self-efficacy	3.46	.72	3.04	.79	3.25	463	.78	35.56***
Intrinsic value	3.87	.69	3.54	.77	3.70	463	.75	23.69***
Extrinsic value	3.95	.77	3.81	.75	3.88	463	.76	3.86*
Intrinsic parental value	3.33	.81	3.06	.75	3.20	458	.79	12.97***
Extrinsic parental value	3.60	.82	3.45	.79	3.52	458	.81	4.20*
Self achievement expectation	2.61	.87	2.06	.93	2.33	457	.94	42.07***
Achievement expectation from parents	2.62	.89	2.20	1.06	2.41	457	1.00	21.76***

Note. *p < .05 ***p < .001

- *Self achievement expectation and achievement expectation from parents by achievement*

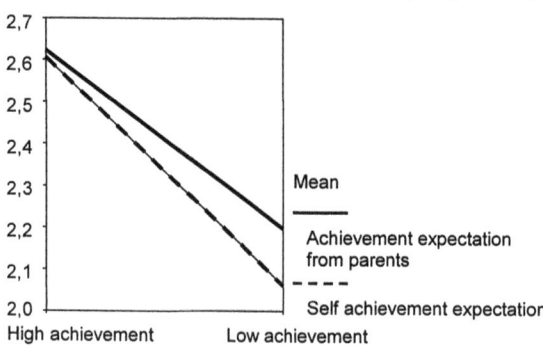

Figure 4.2.1: Self Achievement Expectation and Achievement Expectation from Parents of High and Low Achievement Samples

Self achievement expectation of high achievement sample (M = 2.61, SD = .87) was significantly higher than that (M = 2.06, SD = .93) of low achievement sample (F (1, 455) = 42.07, p < .001). Achievement expectation from parents of high achievement sample (M = 2.62, SD = .89) was significantly higher than that (M = 2.20, SD = 1.06) of low achievement sample (F (1, 455) = 21.76, p < .001).

From the figure of difference line (*Figure 4.2.1*), it is observable that achievement expectation from parents' line is above the self achievement expectation line for both high and low achievement groups.

MANCOVA of the variables by gender controlling prior mathematics achievement

Table 4.2.2: Results of Multivariate Analyses of Variance and Covariance with Gender as Predictor, Mathematics Grade as Covariate, and Attitude toward Math, Self, Values, and Achievement Expectations as Dependent Variables

	Results from MANOVA					Results from MANCOVA					
	Male		Female			Male		Female			
	Mean	*SD*	*Mean*	*SD*	*F*	*Mean*	*SD*	*Mean*	*SD*	*F*	*Partial* η^2
Attitude toward math	3.62	.94	3.34	.90	10.60***	3.65	.90	3.34	.89	43.95***	.20
Math self-concept	3.04	.74	2.78	.64	16.56***	3.01	.72	2.75	.62	37.23***	.17
Self-efficacy	3.44	.79	3.04	.72	31.36***	3.43	.80	3.01	.70	50.11***	.22
Intrinsic value	3.74	.75	3.66	.73	1.32	3.75	.75	3.68	.74	21.38***	.11
Extrinsic value	3.84	.80	3.92	.71	1.40	3.81	.79	3.94	.71	4.62**	.03
Intrinsic parental value	3.16	.87	3.23	.71	1.09	3.13	.87	3.23	.73	12.16***	.06
Extrinsic parental value	3.50	.85	3.54	.76	.23	3.50	.86	3.55	.79	2.11	.01
Self achievement expectation	2.45	1.01	2.21	.84	7.84**	2.50	1.01	2.18	.84	22.65***	.11
Parental achievement expectation	2.64	1.05	2.18	.88	25.41***	2.71	1.06	2.12	.85	24.25***	.12

Note. Sample size for MANOVA: male = 228, female = 223, between group df = 1; sample size for MANCOVA: male = 181, female = 177, between group df = 2; (Variant sample size was due to mathematics grades not being given by some participants.). *p < .05 **p < .01 ***p < .001
Achievement expectation: 1 Bachelor 2 Master 3 Doctor 4 Post-doctor

It is necessary to control prior mathematics achievement since it might confound with gender to predict control-value beliefs, and expectations, etc.. Concerning control-value beliefs, and expectations, a picture favoring male participants emerged. Male participants had more positive attitude toward mathematics, math self-concept, self-efficacy, and higher intrinsic value than female participants. Male participants also had higher level of expectation for themselves and higher expectation from parents than female participants. Female participants reported higher levels of extrinsic value and intrinsic parental value. However, the effect sizes for extrinsic value and intrinsic parental value were too small and therefore, the gender differences cannot be regarded as substantial.

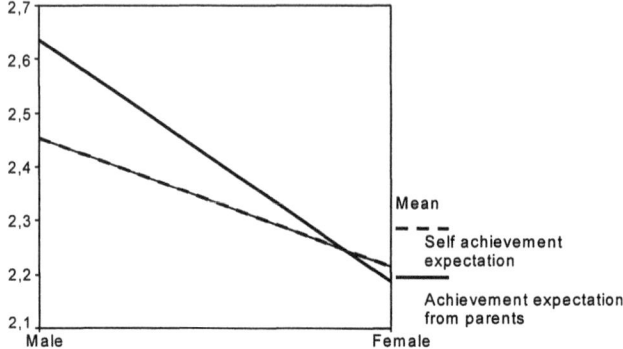

Figure 4.2.2: Difference Lines of Self Achievement Expectation and Achievement Expectation from Parents for Female and Male Samples

From the figure of difference line (*Figure 4.2.2*), it is observable that achievement expectation from parent's line is generally above the self achievement expectation line for both male group and female group. In addition, for the male sample the mean of achievement expectation from parents is above the mean of self achievement expectation; however, for the female sample the mean of achievement expectation from parents is below the mean of self achievement expectation.

MANOVA of the variables by achievement and gender

MANOVA statistics (*Table 4.2.3*) indicated that achievement by gender models of *attitude, self, intrinsic values,* and *achievement expectancy* were all significantly different except for extrinsic value.

- *MANOVA of attitude toward math*

MANOVA of attitude toward math indicated that it was significantly different among achievement by gender groups ($F_{Attitude} = 15.35$ at .001 level, effect size partial $\eta^2 = .09$).

- *MANOVA of math self-concept and self-efficacy*

MANOVAs of self-concept and self-efficacy indicated that both were significantly different among achievement by gender groups ($F_{Self-concept} = 15.33$, $\eta^2 = .09$; $F_{Self-efficacy} = 21.30$, $\eta^2 = .13$, at .001 level respectively).

- *MANOVA of intrinsic value and extrinsic value*

MANOVA of intrinsic value indicated that it was significantly different among achievement by gender groups ($F_{Intrinsic}$ = 8.79 at .001 level, η^2 = .06,). The effect size was rather weak reflecting weak association between intrinsic value and the two independent variables of achievement and gender. MANOVA of extrinsic value indicated that it was not significantly different among achievement by gender groups. In addition, the extrinsic value scores of the four achievement by gender groups were all rather high and above 3.72. The high extrinsic values of the four groups indicated that the four groups viewed math as an important subject for their future no matter whether they liked it or not.

Table 4.2.3: Means and Standard Deviations of Attitude toward Math, Self, Values, and Achievement Expectations by Achievement and Gender

		High achievement group			Low achievement group			df	F (3, 447)	Partial η^2
		Mean	*N*	*SD*	*Mean*	*N*	*SD*			
Attitude toward Math	M	3.76	137	.84	3.42	91	1.04			
	F	3.71	92	.78	3.08	131	.89	3	15.35***	.09
Math self-concept	M	3.14	137	.69	2.90	91	.78			
	F	3.03	92	.57	2.60	131	.62	3	15.33***	.09
Self-efficacy	M	3.54	137	.75	3.30	91	.82			
	F	3.33	92	.66	2.84	131	.70	3	21.30***	.13
Intrinsic value	M	3.83	137	.74	3.61	91	.76			
	F	3.92	92	.62	3.48	131	.75	3	8.79***	.06
Extrinsic value	M	3.92	137	.79	3.72	91	.81			
	F	4.00	92	.73	3.87	131	.69	3	2.29	.02
Intrinsic parental value	M	3.23	137	.87	3.04	91	.86			
	F	3.47	92	.70	3.07	131	.68	3	6.11***	.04
Extrinsic parental value	M	3.50	137	.84	3.51	91	.88			
	F	3.76	92	.77	3.39	131	.72	3	3.88**	.03
Self achievement expectation	M	2.65	137	.96	2.15	91	1.02			
	F	2.54	92	.72	1.97	131	.84	3	15.85***	.10
Achievement expectation from parents	M	2.76	137	.98	2.46	91	1.13			
	F	2.42	92	.70	2.01	131	.96	3	13.90***	.09

Note. **p < .01 ***p < .001 M: Male F: Female
Achievement Expectation: 1 Bachelor 2 Master 3 Doctor 4 Post-doctor

- *MANOVA of parental values*

MANOVAs of parental values indicated that intrinsic parental value and extrinsic parental value were significantly different among achievement by gender groups ($F_{Intrinsic\ parental\ value}$ = 6.11 at .001 level, η^2 = .04; $F_{Extrinsic\ parental\ value}$ = 3.88 at .01 level, η^2 = .03). The effect sizes were rather weak reflecting weak associations between parental values and the two independent variables of achievement and gender.

- *MANOVA of achievement expectations*

MANOVAs of expectations indicated that self achievement expectation and achievement expectation from parents were significantly different among achievement by gender groups ($F_{\text{Self achievement expectancy}} = 15.85$, $\eta^2 = .10$; $F_{\text{Achievement expectany from parents}} = 13.90$, $\eta^2 = .09$, at .001 level respectively). Although the effect sizes were low, the chances of finding significant effects in the current study were enough.

- *Achievement values and expectations between child and parents*

Besides multivariate analyses of variance, there was an important issue to investigate, namely consistency of achievement values and expectations between child and parents. Therefore, zero – order correlations of achievement values and expectations between child and parents were conducted. Pearson correlations between child and parents for *intrinsic value*, *extrinsic value*, and *achievement expectation* were: .51, .21, and .64, at .01 level (2-tailed). The moderate Pearson correlations implied that child and parents shared somewhat similar achievement values and expectations in learning mathematics. An abrupt generation gap on achievement values and expectations did not seem to exist.

Table 4.2.4: Zero – Order Correlations of Achievement Values and Expectations between Child and Parents

	1	2	3	4	5	6
1 Intrinsic value						
2 Extrinsic value	.36**					
3 Parental intrinsic value	.51**	.24**				
4 Parental extrinsic value	.30**	.21**	.50**			
5 Self achievement expectation	.29**	.16**	.17**	.08		
6 Achievement expectation from parents	.18**	.08	.07	.09	.64**	
Mean	3.70	3.88	3.20	3.52	2.33	2.41
SD	.75	.76	.79	.81	.94	1.00
N	463	463	458	458	457	457

*Note. ** Correlation is significant at the .01 level (2- tailed).*

Post hoc multiple comparisons of the variables by achievement and gender

Achievement by gender grouping condition decided the four groups, namely (1) High achievement male group, (2) High achievement female group, (3) Low achievement male group, (4) Low achievement female group. Tukey test, one of the Post Hoc tests was applied to identify among and between group differences. (*Table 4.2.5*).

- *Attitude toward math, self, achievement values, and achievement expectancies by achievement and gender*

Figure 4.2.3 showed that high achievers (both male and female) had more positive attitude toward math than low achievers. Low achievement females had the lowest math self-concept and self-efficacy among the four groups. There were not many differences concerning achievement values among the four groups except that high achievement females' intrinsic value, intrinsic parental value and extrinsic parental value were higher than those of low achievement females.

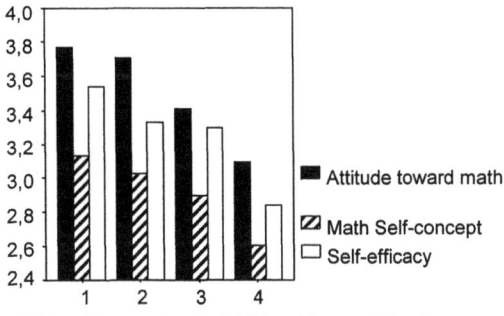

1 High achievement male   2 High achievement female
3 Low achievement male   4 Low achievement female

Figure 4.2.3: Self and Attitude towards Math by Achievement and Gender

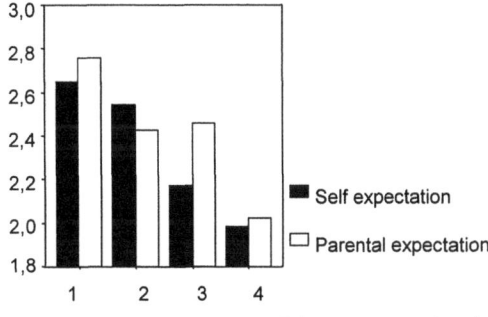

1 High achievement male   2 High achievement female
3 Low achievement male   4 Low achievement female

Figure 4.2.4: Achievement Expectations by Achievement and Gender

The bar graph (*Figure 4.2.4*) showed that self-expectation was consistent with achievement level, whereas parental expectation for low achievement males was even above parental expectation for high achievement females.

- *High achievement males vs. high achievement females*

According to the Tukey HSD test, high achievement males and high achievement females did not differ on any of attitude toward math, math self-concept, and self-efficacy in math learning, achievement values, parental values, or achievement expectations.

Table 4.2.5: Tukey Test of Attitude, Self, Values, and Achievement Expectations by Achievement and Gender

	Achievement By gender (I)	Achievement By gender (J)	Mean Difference (I-J)	Std. Error	Sig.
Attitude, Self, Values, and Expectations					
Attitude toward	HM	HF	6.16E-02	.12	.955
Math	HF	LF	.62*	.12	.000
	LM	HM	-.35*	.12	.015
	LF	LM	-.33	.12	.032
Math self-concept	HM	HF	.11	8.98E-02	.646
	HF	LF	.42*	9.06E-02	.000
	LM	HM	-.23	8.87E-02	.053
	LF	LM	-.30*	8.95E-02	.004
Self-efficacy	HM	HF	.21	9.86E-02	.146
	HF	LF	.49*	9.94E-02	.000
	LM	HM	-.23	9.73E-02	.077
	LF	LM	-.47*	9.82E-02	.000
Intrinsic value	HM	HF	-8.84E-02	9.85E-02	.806
	HF	LF	.44*	9.94E-02	.000
	LM	HM	-.24	9.72E-02	.059
	LF	LM	-.11	9.81E-02	.707
Extrinsic value	HM	HF	-9.64E-02	.10	.782
	HF	LF	.13	.10	.573
	LM	HM	-.20	.10	.189
	LF	LM	.17	.10	.365
Intrinsic parental	HM	HF	-.24	.11	.113
value	HF	LF	.40*	.11	.001
	LM	HM	-.18	.10	.319
	LF	LM	1.55E-02	.11	.999
Extrinsic parental	HM	HF	-.26	.11	.074
value	HF	LF	.37*	.11	.004
	LM	HM	9.51E-03	.11	1.000
	LF	LM	-.12	.11	.695
Self achievement	HM	HF	.11	.12	.818
expectation	HF	LF	.56*	.12	.000
	LM	HM	-.48*	.12	.000
	LF	LM	-.19	.12	.422
Achievement	HM	HF	.34	.13	.048
expectation from	HF	LF	.40*	.13	.012
parents	LM	HM	-.30	.13	.089
	LF	LM	-.43*	.13	.005

Note: The mean difference is significant at the .05 level.
HM: High achievement male   HF: High achievement female
LM: Low achievement male   LF: Low achievement female

- *High achievement females vs. low achievement females*

According to the Tukey HSD test, except for extrinsic value high achievement females and low achievement females differed on attitude toward math (Mean difference = .62*), math self-concept (Mean difference = .42*), self-efficacy in learning math (Mean difference = .49*), intrinsic achievement value (Mean difference = .44*), intrinsic parental value (Mean difference =. 40*), extrinsic parental value = .37*), self-expectation (Mean difference = .56*), and parental expectation (Mean difference = .40*). The statistics favored high achievement females.

- Low achievement females vs. low achievement males

According to the Tukey HSD test, low achievement females had less positive math self-concept and self-efficacy in math learning than low achievement males (Mean difference = -.30*, -.47*, respectively). Although low achievement females and low achievement males did not differ on attitude toward math, achievement values, parental values, or self achievement expectation, parents of low achievement females had lower achievement expectation for them than parents of low achievement males (Mean difference = -.43*).

- Low achievement males vs. high achievement males

According to the Tukey HSD test, low achievement males and high achievement males differed on attitude toward math and self-expectation of degree attainment (Mean = -.35*, -.48*, respectively). It is interesting to note that they did not differ on math self-concept, self-efficacy in math learning, achievement values, perceived parental values, or achievement expectation from parents.

Summary: Among group differences of attitude, self, values, and expectancies by achievement and gender

Among group differences of attitude, self, values, and achievement expectations were more frequently found by achievement than by gender. For example, high achievement female and low achievement female differed on all of the above variables except for extrinsic value. Gender impact was only found on low achievement female and low achievement male, who differed from each other on math self-concept, self-efficacy, and achievement expectation from parents. Gender impact was not found on high achievement female and high achievement male.

Although males were found to have more positive attitude toward mathematics and intrinsic value than females when controlling their prior math achievement (*Table 4.2.2*), the results based on the Post hoc multiple comparisons were not in contradiction with MANCOVA results due to the fact that the MANCOVA comparisons were between male and female two groups while Post hoc multiple comparisons were among achievement by gender four groups. The four group comparisons are more beneficial for us to see among group differences.

Although the effect sizes of the among group differences of the major variables were low, a number of investigators have argued that even extremely small effects may have significant practical implications (Abelson, 1985; Martell, Lane, & Emrich, 1996). The effect size is also a function of the sample size. Therefore, increased effect sizes could be expected when the sample size is larger.

4.2.2 Academic emotions, learning strategies, and achievement goals

MANOVA of the variables by achievement

- *MANOVA statistics of academic emotions, learning strategies, and achievement goals of high and low achievement samples*

Based on the MANOVA statistics, a clear pattern of math learning processes and achievement has emerged. High achievement group and low achievement group differed significantly on the affective-cognitive processes in math learning expect for experiencing shame, using rehearsal strategy, and taking mastery avoidance goal.

Academic emotions by achievement Concerning affective processes, high achievement group experienced more enjoyment (Mean $_H$ = 3.46, Mean $_L$ = 2.96, F (1, 458) = 67.00, p < .001) and pride (Mean $_H$ = 3.54, Mean $_L$ = 3.12, F (1, 455) = 37.20, p < .001), and less anger (Mean $_H$ = 1.86, Mean $_L$ = 2.51, F (1, 458) = 103.11, p < .001), anxiety (Mean $_H$ = 2.35, Mean $_L$ = 2.73, F (1, 457) = 39.92, p < .001), hopelessness (Mean $_H$ = 2.26, Mean $_L$ = 2.79, F (1, 455) = 49.27, p < .001), and boredom (Mean $_H$ = 1.88, Mean $_L$ = 2.68, F (1, 458) = 133.12, p < .001) than low achievement group. High achievement group and Low achievement group did not differ in experiencing shame.

Table 4.2.6: **Means and Standard Deviations of Academic Emotions, Learning Strategies, and Achievement Goals of High and Low Achievement Samples**

	High achievement group			Low achievement Group				
	N	*Mean*	*SD*	*N*	*Mean*	*SD*	*df*	*F*
Achievement emotions								
Enjoyment	231	3.46	.60	229	2.96	.70	1	67.00***
Pride	230	3.54	.72	227	3.12	.74	1	37.20***
Anger	231	1.86	.56	229	2.51	.79	1	103.11***
Anxiety	230	2.35	.61	229	2.73	.69	1	39.92***
Shame	230	2.51	.59	227	2.57	.61	1	.96
Hopelessness	230	2.26	.72	227	2.79	.88	1	49.27***
Boredom	231	1.88	.61	229	2.68	.87	1	133.12***
Learning strategies								
Rehearsal	231	3.38	1.08	230	3.32	1.19	1	.34
Elaboration	231	4.10	1.06	229	3.71	1.23	1	13.69***
Organization	231	3.73	1.15	230	3.30	1.21	1	15.38***
Critical thinking	231	4.73	1.18	230	3.86	1.38	1	52.18***
Metacognitive self-regulation	231	4.77	.81	228	4.05	1.01	1	69.69***
Achievement goals								
Performance approach	231	5.14	1.42	229	4.58	1.59	1	16.05***
Mastery avoidance	231	5.08	1.51	229	5.16	1.49	1	.36
Mastery approach	231	6.31	.81	229	5.65	1.29	1	42.96***
Performance avoidance	231	3.70	1.53	229	4.59	1.40	1	42.47***

Note: ***p < .001

Learning strategies by achievement Concerning learning strategies, high achievement group used more elaboration (Mean $_H$ = 4.10, Mean $_L$ = 3.71, F (1, 458) = 13.69, p < .001), organization (Mean $_H$ = 3.73, Mean $_L$ = 3.30, F (1, 459) = 15.38, p < .001), critical thinking (Mean $_H$ = 4.73, Mean $_L$ = 3.86, F (1, 459) = 52.18, p < .001), and metacognitive self-regulation strategies (Mean $_H$ = 4.77, Mean $_L$ = 4.05, F (1, 457) = 69.69, p < .001) than low achievement group. High achievement group and low achievement group did not differ in applying rehearsal strategy.

Achievement goals by achievement Concerning achievement motivation, high achievement group took more approach goals, namely performance approach (Mean $_H$ = 5.14, Mean $_L$ = 4.58, F (1, 458) = 16.05, p < .001) and mastery approach (Mean $_H$ = 6.31, Mean $_L$ = 5.65, F (1, 458) = 42.96, p < .001) and less performance avoidance goal (Mean $_H$ = 3.70, Mean $_L$ = 4.59, F (1, 458) = 42.47, p < .001) than low achievement group. High achievement group and low achievement group did not differ on mastery avoidance goal.

Summary In conclusion, high achievement group experienced more positive affect and less negative affect in math learning than low achievement group. In addition, they applied more cognitive and metacognitive strategies in math learning process and took more approach goals and less avoidance goal than low achievement group.

MANCOVA of the variables by gender when controlling prior math achievement

- *MANOVA and MANCOVA statistics of academic emotions, learning strategies, and achievement goals*

Gender is a sensitive issue when math learning processes are explored. In addition, prior math achievement might confound with gender to predict emotions, strategies, and goals. Therefore it was necessary to analyze math learning processes by gender when controlling prior math achievement.

Achievement emotions by gender Concerning achievement emotions, a picture favoring male adolescents emerged. Male adolescents had higher levels of enjoyment and pride, and lower levels of anger, anxiety, hopelessness, and boredom than female adolescents. Male participants reported less shame, too. However, the effect size for shame was too small and therefore, the gender difference cannot be regarded as substantial.

Learning strategies by gender Concerning learning strategies, male adolescents were found to apply more critical thinking. metacognitive self-regulation, and elaboration strategies than female adolescents. Male and female participants differed on rehearsal, and organization according to their reported scores. However, the effect sizes were too small and therefore, the gender differences cannot be regarded as substantial.

Achievement goals by gender Female participants reported more mastery approach and performance avoidance goals than male participants. Males reported more performance approach than females. However, the effect size was much smaller and therefore, the gender difference cannot be regarded as substantial.

Summary A short summary based on the above results describes such a picture: male participants experiencing more positive affect and less negative affect in math learning process than female participants; using more critical thinking, elaboration, and metacognitive self-regulation strategies than female participants; applying less performance avoidance goal than female participants. Female participants applied more mastery approach goal than male participants. However, male and female participants did not differ on simple cognitive learning strategies, performance approach and mastery avoidance goal. It is worthwhile to mention that it is necessary to control prior mathematics achievement since a few more significant results emerged when using gender to predict emotions, learning strategies, and achievement goals when controlling prior math achievement.

Table 4.2.7: **Results of Multivariate Analyses of Variance and Covariance with Gender as Predictor, Mathematics Grade as Covariate, and Emotions, Learning Strategies, and Achievement Goals as Dependent Variables**

	Results from MANOVA					Results from MANCOVA					
	Male		Female			Male		Female			
	Mean	*SD*	*Mean*	*SD*	*F*	*Mean*	*SD*	*Mean*	*SD*	*F*	*Partial* η^2
Achievement emotions											
Enjoyment	3.28	.72	3.12	.68	5.87*	3.30	.70	3.11	.68	57.09***	.24
Pride	3.38	.78	3.30	.73	1.32	3.38	.74	3.29	.75	28.29***	.14
Anger	2.10	.73	2.27	.78	5.99*	2.04	.70	2.25	.78	56.96***	.24
Anxiety	2.44	.68	2.64	.67	10.10**	2.40	.67	2.63	.67	39.74***	.18
Shame	2.52	.59	2.55	.61	.30	2.49	.57	2.53	.63	7.50***	.04
Hopelessness	2.42	.84	2.64	.84	7.56**	2.37	.82	2.63	.83	50.08***	.22
Boredom	2.20	.81	2.36	.89	3.84	2.16	.79	2.35	.92	76.08***	.30
Learning strategies											
Rehearsal	3.27	1.12	3.43	1.15	2.33	3.19	1.05	3.43	1.12	3.59*	.02
Elaboration	4.00	1.16	3.83	1.18	2.53	3.95	1.12	3.81	1.20	18.46***	.09
Organization	3.42	1.22	3.63	1.18	3.33	3.32	1.18	3.63	1.18	11.95***	.06
Critical thinking	4.61	1.29	3.99	1.36	24.50***	4.54	1.29	3.97	1.39	40.97***	.19
Metacognitive self-regulation	4.57	.96	4.25	.99	12.13***	4.57	.95	4.24	.99	55.60***	.24
Achievement goals											
Performance approach	4.87	1.46	4.83	1.61	.09	4.86	1.46	4.84	1.58	6.57**	.04
Mastery avoidance	5.01	1.39	5.22	1.62	2.15	4.99	1.37	5.26	1.57	1.49	.01
Mastery approach	5.90	1.10	6.07	1.13	2.63	5.93	1.06	6.11	1.13	23.57***	.12
Performance avoidance	3.84	1.52	4.44	1.52	17.44***	3.70	1.47	4.46	1.52	25.14***	.12

Note. Sample size for MANOVA: male = 229, female = 221, between group df = 1; sample size for MANCOVA: male = 182, female = 176, between group df = 2; (Variant sample size was due to mathematics grades not being given by some participants.). *p < .05   **p < .01   ***p < .001

MANOVA of the variables by achievement and gender

- *Academic emotions by achievement and gender*

Table 4.2.8: Means and Standard Deviations of Achievement Emotions, Learning Strategies, and Achievement Goals by Achievement and Gender

		High achievement group			Low achievement group					
		Mean	N	SD	Mean	N	SD	df	F(3, 446)	Partial η²
Achievement Emotions										
Enjoyment	M	3.45	138	.63	3.04	91	.77			
	F	3.47	91	.57	2.88	130	.65	3	23.39***	.14
Pride	M	3.50	138	.76	3.19	91	.77			
	F	3.60	91	.64	3.08	130	.72	3	12.84***	.08
Anger	M	1.90	138	.58	2.40	91	.82			
	F	1.81	91	.52	2.59	130	.77	3	35.63***	.20
Anxiety	M	2.32	138	.64	2.63	91	.70			
	F	2.39	91	.56	2.82	130	.69	3	15.27***	.09
Shame	M	2.51	138	.60	2.53	91	.58			
	F	2.51	91	.58	2.56	130	.61	3	.38	.003
Hopelessness	M	2.22	138	.75	2.71	91	.89			
	F	2.33	91	.69	2.85	130	.87	3	17.03***	.10
Boredom	M	1.93	138	.63	2.61	91	.88			
	F	1.79	91	.57	2.76	130	.86	3	46.21***	.24
Learning strategies										
Rehearsal	M	3.20	138	1.08	3.36	91	1.17			
	F	3.64	91	1.05	3.28	130	1.20	3	2.86*	.02
Elaboration	M	4.10	138	1.11	3.84	91	1.21			
	F	4.13	91	.99	3.61	130	1.26	3	5.21**	.03
Organization	M	3.53	138	1.17	3.25	91	1.27			
	F	4.04	91	1.08	3.34	130	1.17	3	8.99***	.06
Critical thinking	M	4.87	138	1.18	4.20	91	1.35			
	F	4.52	91	1.17	3.62	130	1.36	3	23.66***	.13
Metacognitive Self-regulation	M	4.79	138	.85	4.25	91	1.03			
	F	4.74	91	.77	3.91	130	.99	3	25.49***	.15
Achievement goals										
Performance approach	M	5.10	138	1.40	4.52	91	1.48			
	F	5.17	91	1.45	4.59	130	1.67	3	5.34***	.03
Mastery avoidance	M	5.09	138	1.39	4.89	91	1.39			
	F	5.05	91	1.69	5.34	130	1.56	3	1.83	.01
Mastery approach	M	6.19	138	.86	5.48	91	1.27			
	F	6.48	91	.69	5.79	130	1.28	3	17.47***	.10
Performance avoidance	M	3.56	138	1.51	4.26	91	1.44			
	F	3.91	91	1.56	4.81	130	1.38	3	17.73***	.10

Note. *p < .05 **p < .01 ***p < .001

Cognitive-affective differences by gender were found. A further step was to observe the participants in their ability groups when their gender was considered, too. MANOVA of achievement emotions indicated that achievement emotions were significantly different among achievement by gender group except for experiencing shame ($F_{Enjoyment}$ (3, 446) =23.39, with effect size partial η^2 = .14; F_{Pride} (3, 446) = 12.84, η^2 = .08; F_{Anger} (3, 446) = 35.63, η^2 = .20; $F_{Anxiety}$ (3, 446) = 15.27, η^2 = .09; $F_{Hopelessness}$ (3, 446) = 17.03, η^2 = .10; $F_{Boredom}$ (3, 446) = 46.21, η^2 = .24, at .001 level; *Table 4.2.8*).

- *Learning strategies by achievement and gender*

MANOVA of learning strategies indicated that learning strategies were all significantly different among achievement by gender group ($F_{Rehearsal}$ (3, 446) = 2.86 at 0.05 level, η^2 = .02; $F_{Elaboration}$ (3, 446) = 5.21 at 0.01 level, η^2 = .03; $F_{Organization}$ (3, 446) = 8.99 at 0.001 level, η^2 = .06; $F_{Critical\ thinking}$ (3, 446) = 23.66 at 0.001 level, η^2 = .13; $F_{Metacognitive}$ (3, 446) = 25.49 at 0.001 level, η^2 = .15; *Table 4.2.8*). However, the effect sizes for rehearsal, elaboration, and organization strategies were weak reflecting weak associations between the learning strategies and the two independent variables of achievement and gender.

- *Achievement goals by achievement and gender*

MANOVA of achievement goals indicated that achievement goals were significantly different among achievement by gender group except for mastery avoidance goal ($F_{Performance\ approach}$ (3, 446) = 5.34, η^2 = .03; $F_{Mastery\ approach}$ (3, 446) = 17.47, η^2 = .10; $F_{Performance\ avoidance}$ (3, 446) = 17.73 ,η^2 = .10 at 0.001 level, *Table 4.2.8*). However, the effect size of performance approach goal was weak reflecting weak associations between the performance approach goal and the two independent variables of achievement and gender. The effect size is also a function of the sample size. Therefore, increased effect sizes could be expected when the sample size is larger.

Post hoc multiple comparisons of the variables by achievement and gender

Table 4.2.9: Tukey Test of Achievement Emotions, Learning Strategies, and Achievement Goals by Achievement and Gender

	Achievement by gender (I)	Achievement by gender (J)	Mean Difference (I-J)	Std. Error	Sig.
Achievement emotions					
Enjoyment	HM	HF	-2.50E-02	8.82E-02	.992
	HF	LF	.59*	8.90E-02	.000
	LM	HM	-.39**	8.77E-02	.000
	LF	LM	-.18	8.85E-02	.196
Pride	HM	HF	-9.06E-02	9.81E-02	.793
	HF	LF	.51*	9.92E-02	.000
	LM	HM	-.31*	9.78E-02	.008
	LF	LM	-.11	9.89E-02	.660
Anger	HM	HF	8.66E-02	9.20E-02	.783
	HF	LF	-.78*	9.28E-02	.000
	LM	HM	.50*	9.14E-02	.000
	LF	LM	.19	9.22E-02	.149
Anxiety	HM	HF	-7.60E-02	8.81E-02	.824
	HF	LF	-.42*	8.89E-02	.000
	LM	HM	.30*	8.72E-02	.004
	LF	LM	.20	8.80E-02	.097
Shame	HM	HF	-1.81E-02	8.13E-02	1.00
	HF	LF	-6.94E-02	8.21E-02	.833
	LM	HM	2.72E-02	8.10E-02	.987
	LF	LM	4.40E-02	8.19E-02	.950

Hopelessness	HM	HF	-.10	.11	.773
	HF	LF	-.53*	.11	.000
	LM	HM	.48*	.11	.000
	LF	LM	.15	.11	.496
Boredom	HM	HF	.14	.10	.491
	HF	LF	-.97*	.10	.000
	LM	HM	.65*	.10	.000
	LF	LM	.17	.10	.314
Learning strategies					
Rehearsal	HM	HF	-.43*	.15	.025
	HF	LF	.36	.15	.094
	LM	HM	.13	.15	.819
	LF	LM	-5.92E-02	.15	.980
Elaboration	HM	HF	-9.66E-02	.15	1.000
	HF	LF	.49*	.16	.009
	LM	HM	-.27*	.15	.275
	LF	LM	-.21	.15	.526
Organization	HM	HF	-.50*	.16	.008
	HF	LF	.69*	.16	.000
	LM	HM	-.31	.16	.182
	LF	LM	.12	.16	.863
Critical thinking	HM	HF	.36	.17	.157
	HF	LF	.91*	.17	.000
	LM	HM	-.67*	.17	.000
	LF	LM	-.60*	.17	.003
Metacognitive	HM	HF	4.59E-02	.12	.982
Self-regulation	HF	LF	.82*	.12	.000
	LM	HM	-.55*	.12	.000
	LF	LM	-.32	.12	.045
Achievement goals					
Performance approach	HM	HF	-7.85E-02	.20	.981
	HF	LF	.62*	.21	.015
	LM	HM	-.53	.20	.044
	LF	LM	-1.06E-02	.20	1.000
Mastery avoidance	HM	HF	2.30E-02	.20	.999
	HF	LF	-.28	.20	.502
	LM	HM	-.20	.20	.759
	LF	LM	.46	.20	.105
Mastery approach	HM	HF	-.30	.14	.152
	HF	LF	.69*	.14	.000
	LM	HM	-.72*	.14	.000
	LF	LM	.33	.14	.093
Performance avoidance	HM	HF	-.34	.20	.309
	HF	LF	-.91*	.20	.000
	LM	HM	.73*	.19	.001
	LF	LM	.53	.20	.035

Note. * The mean difference is significant at the .05 level.　　HM: High achievement male
HF: High achievement female　　LM: Low achievement male　　LF: Low achievement female

Achievement by gender grouping condition decided the following four groups, namely (1) high achievement male, (2) high achievement female, (3) low achievement male, (4) low achievement female. In order to identify among and between group differences, Tukey test, one of the Post Hoc tests was applied to achieve this goal (*Table 4.2.9*).

- *Achievement emotions by achievement and gender*

Concerning achievement emotions, positive emotions (enjoyment and pride) and negative emotions (anger, anxiety, hopelessness, and boredom) were found to significantly differ at achievement level rather than gender category. High achievement females had more positive emotions and less negative emotions than low achievement females (Mean difference $_{Enjoyment}$ = .59*, Mean difference $_{Pride}$ = .51*, Mean difference $_{Anger}$ = -.78*, Mean difference $_{Anxiety}$ = -.42*, Mean difference $_{Hopelessness}$ = -.53*, Mean difference $_{Boredom}$ = -.97* at .05 level). Low achievement males had less positive emotions and more negative emotions than high achievement males (Mean difference $_{Enjoyment}$ = -.39*, Mean difference $_{Pride}$ = -.31*, Mean difference $_{Anger}$ = .50*, Mean difference $_{Anxiety}$ = .30*, Mean difference $_{Hopelessness}$ = .48*, Mean difference $_{Boredom}$ = .65*). Gender differences for experiencing academic emotions were not found. Differences on shame were not found either at achievement level or gender category.

- *Learning strategies by achievement and gender*

A more complex picture emerged concerning learning strategies. Significant mean differences were found at both achievement level and gender category. High achievement males used less rehearsal strategies than high achievement females (Mean difference = -.43* at .05 level). High achievers (females and males) used more elaboration strategies than their counterparts (Mean difference = .49*, .27* respectively). High achievement males used less organization strategies than high achievement females (Mean difference = -.50*). High achievement females used more organization strategies than low achievement females (Mean difference = .69*). High achievers (females and males) used more critical thinking strategies than their counterparts (Mean difference = .91*, .67*). Low achievement females used less critical thinking strategies than low achievement males (Mean difference = -.60*). High achievers (females and males) used more metacognitive self-regulation strategies than their counterparts (Mean difference = .82*, .55*).

- *Achievement goals by achievement and gender*

Differences for taking achievement goals were only found by achievement. High achievement females had higher level of performance approach than low achievement females (Mean difference = .62* at .05 level). High achievers (both females and males) had more mastery approach than their counterparts (Mean difference = .69*, .72*). High achievers (both females and males) had less performance avoidance goal than their counterparts (Mean difference = -.91*, -.73*). Differences on mastery avoidance were not found. However, Kenney-Benson, Pomerantz, Ryan, and Patrick (2006) found that girls were more likely than boys to hold mastery over performance goals. The researchers did not consider their sample's prior achievement level.

Summary of emotions, learning strategies and goals by achievement and gender

Based on the above statistic analysis, a short conclusion and prediction were made. Differences on achievement emotions, and achievement goals were found by achievement rather than by gender. Differences on learning strategy use were found by both achievement and gender. Therefore, it was meaningful to use gender as a potential predictor to predict learning strategies but not as a predictor to predict achievement emotions.

Although males were found to have more positive affect and less negative affect than females when controlling their prior math achievement (*Table 4.2.7*), the results based on the Post hoc multiple comparisons were not in contradiction with MANCOVA results due to the fact that the MANCOVA comparisons were between male and female two groups while Post hoc multiple comparisons were among achievement by gender four groups. The four group comparisons are more beneficial for us to see among group differences.

Although the effect sizes of the among group differences of the major variables were low, a number of investigators have argued that even extremely small effects may have significant practical implications (Abelson, 1985; Martell, Lane, & Emrich, 1996). The effect size is also a function of the sample size. Therefore, increased effect sizes could be expected when the sample size is larger.

4.2.3 Summary of all the major variable comparisons by achievement and gender

Since it was the researcher's purpose to compare two different achievement groups and at the mean time gender was a sensitive issue concerning math learning processes and outcome, therefore achievement by gender four group comparisons were conducted by MANOVA and Tukey test. The table (*Table 4.2.10*) summarized the among group differences for the major variables, namely, attitude toward math, self, values, expectations, academic emotions, learning strategies, and achievement goals.

A clear pattern concerning the major variables has been identified: attitude, values, self achievement expectation, academic emotions, and achievement goals only differed at achievement level; math self-concept, self-efficacy in math learning, achievement expectation from parents, and learning strategies differed at both achievement level and gender category.

- *High achievement females vs. low achievement females*

Specifically speaking, the most between group differences were found between high achievement females and low achievement females with positive side favoring high achievement females. They differed on attitude toward math, math self-concept, self-efficacy, intrinsic value, perceived parental values, achievement expectancies, academic emotions except for shame, learning strategies except for rehearsal, and achievement goals except for mastery avoidance goal.

- *Low achievement males vs. high achievement males*

Table 4.2.10: Among Group Differences of Major Variables by Achievement and Gender

	High female vs. Low female	Low female vs. Low male	Low male vs. High male	High male vs. High female
	Achievement difference	*Gender difference*	*Achievement difference*	*Gender difference*
Attitude toward math	X +		X -	
Self				
Math self-concept	X +	X -		
Self-efficacy	X +	X -		
Values				
Intrinsic value	X +			
Extrinsic value				
Parental intrinsic value	X +			
Parental extrinsic value	X +			
Achievement expectancy				
Self achievement expectancy	X +		X -	
Achievement expectancy from parents	X +	X -		
Academic emotions				
Enjoyment	X +		X -	
Pride	X +		X -	
Anger	X -		X +	
Anxiety	X -		X +	
Shame				
Hopelessness	X -		X +	
Boredom	X -		X +	
Learning strategies				
Rehearsal				X -
Elaboration	X +		X -	
Organization	X +			X -
Critical thinking	X +	X -	X -	
Metacognitive self-regulation	X +		X -	
Achievement goals				
Performance approach	X +			
Mastery avoidance				
Mastery approach	X +		X -	
Performance avoidance	X -		X +	

Note. High female: High achievement female; Low female: Low achievement female
High male: High achievement male; Low male: Low achievement male

The second most between group differences were found between low achievement males and high achievement males with positive side favoring high achievement males. They differed on attitude toward math, self achievement expectancy, and academic emotions except for shame and learning strategies except for rehearsal and organization, and achievement goals except for performance approach and mastery avoidance. It is important to note that low achievement males and high achievement males did not differ on math self-concept, self-efficacy, values, and parental expectation.

- *Low achievement females vs. low achievement males*

The third most between group differences were identified between low achievement females and low achievement males at gender category. They differed on math self-concept, self-efficacy, achievement expectancy from parents, and critical thinking with positive side favoring low achievement males. They did not differ on attitude toward math, self achievement expectancy, academic emotions, learning strategies except for critical thinking, or achievement goals.

- *High achievement males vs. high achievement females*

The least between group differences were identified between high achievement males and high achievement females. High achievement females used more rehearsal and organization strategies than high achievement males. They did not differ on attitude toward math, math self-concept, self-efficacy, values, achievement expectancies, academic emotions, or achievement goals.

- *Summary of all the variables by achievement and gender*

In conclusion, high achievement females and high achievement males were almost identical. They had similar level of math self-concept, self-efficacy, values, self achievement expectation and achievement expectation from parents. In addition, they did not differ on academic emotions, strategy use except for rehearsal and organization, or achievement goals. Low achievement females and low achievement males although they did not differ on attitude toward math, values, self achievement expectancy, academic emotions, learning strategy use expect for critical thinking, or achievement goals, males had more positive math self-concept, higher self-efficacy, and higher achievement expectation from parents than females. Low achievement males' math self-concept, self-efficacy, and perceived achievement expectancy from parents were so high that they did not differ from high achievement males' math self-concept, self-efficacy or achievement expectation from parents. Girls seemed to have to be real high achievers to buffer their math self-concept and self-efficacy in math learning. Parental expectation for girls was only high when they were real high achievers.

By so far, the **Research Question III**: "*Do high achievers' cognition-emotion processes differ from low achievers' processes? Do male students' cognition-emotion processes differ from female students' processes?*" has been well explored. The key to answer the above research question is that learners' cognition-emotion processes should always be considered with the conjunction of learners' achievement level and his or her gender.

4.3 Interrelations between affective-cognitive processes
4.3.1 Intercorrelations among variables in general

Table 4.3.1: Intercorrelations among Variables

	1	2	3	4	5	6	7	8	9	10	11	12	13	14	15	16	17	18	19	20	21	22	23	24	25	26	27	28
1.Enjoyment	-																											
2.Pride	.72	-																										
3.Anger	-.68	-.44	-																									
4.Anxiety	-.55	-.29	.75	-																								
5.Shame	-.24	-.06	.47	.65	-																							
6.Hopelessness	-.55	-.33	.73	.84	.56	-																						
7.Boredom	-.73	-.52	.82	.64	.33	.61	-																					
8.Rehearsal	.37	.37	-.21	-.11	-.002	-.14	-.29	-																				
9.Elaboration	.56	.48	-.43	-.35	-.14	-.36	-.44	.59	-																			
10.Organization	.43	.39	-.31	-.18	-.03	-.22	-.35	.62	.59	-																		
11.Critical thinking	.50	.44	-.44	-.41	-.21	-.38	-.44	.40	.60	.38	-																	
12.Meta SR	.64	.51	-.58	-.49	-.26	-.50	-.62	.49	.72	.52	.68	-																
13.PerApproach	.17	.29	-.07	.12	.17	.10	-.14	.15	.15	.19	.23	.20	-															
14.MasAvoidance	.03	.12	.10	.30	.27	.21	.04	.21	.05	.12	.05	.06	.43	-														
15.MasApproach	.36	.37	-.28	-.12	-.02	-.17	-.32	.26	.29	.28	.33	.44	.48	.55	-													
16.PerAvoidance	-.34	-.18	.41	.46	.35	.41	.40	.05	-.15	-.02	-.22	-.27	.31	.41	.13	-												
17.Self-concept	.50	.43	-.40	-.46	-.31	-.45	-.37	.16	.37	.24	.46	.48	.10	-.16	.13	-.27	-											
18.Self-efficacy	.55	.45	-.46	-.49	-.32	-.47	-.44	.18	.41	.25	.48	.51	.16	-.10	.19	-.30	.77	-										
19.Intrinsic value	.67	.51	-.49	-.37	-.13	-.37	-.51	.25	.40	.36	.41	.50	.23	.09	.37	-.25	.47	.52	-									
20.Extrinsic value	.21	.27	-.06	.10	.17	.07	-.11	-.38	.29	.14	.16	.14	.53	.36	.41	.06	.28	.09	.13	.35	-							
21.Intrinsic PV	.57	.53	-.37	-.29	-.15	-.31	-.38	.29	.37	.31	.40	.42	.17	.06	.29	-.15	.37	.38	.51	.51	-							
22.Extrinsic PV	.28	.30	-.14	-.06	.03	-.05	-.17	.15	.15	.18	.41	.25	.18	.13	.11	.24	-.06	.21	.36	.17	.40	.25	-					
23.Lecture engagement	.72	.48	-.58	-.44	-.13	-.45	-.67	.27	.45	.32	.43	.54	.10	.03	.34	-.25	.40	.45	.61	.17	.40	.25	-					
24.Harsh evaluation	-.31	-.20	.32	.25	.17	.23	.33	.01	-.07	-.04	-.05	-.11	-.06	.02	-.12	.16	-.20	-.24	-.16	.05	-.07	-.02	-.27	-				
25.Evaluation focus	-.42	-.30	.40	.29	.15	.32	.43	-.12	-.25	-.18	-.20	-.29	-.02	-.02	-.21	.22	-.20	-.24	-.34	-.10	-.20	-.12	-.52	.40	-			
26.Parental control	-.24	-.12	.37	.38	.29	.39	.38	-.01	-.12	-.14	-.20	-.31	.07	.13	-.16	.30	-.19	-.21	-.21	.07	-.09	.09	-.27	.27	.26	-		
27.Peer attitude	.47	.42	-.30	-.17	.02	-.22	-.40	.20	.26	.25	.21	.30	.09	.10	.24	-.18	.22	.25	.41	.08	.39	.37	.49	-.18	-.36	-.13	-	
28.Teacher enthusiasm	.46	.31	-.39	-.26	-.07	-.25	-.48	.15	.27	.16	.29	.36	.16	.07	.29	-.15	.20	.25	.33	.16	.24	.20	.66	-.19	-.38	-.24	.43	-

Note. Meta SR = Metacognitive self-regulation. Intrinsic PV = Intrinsic parental value. Extrinsic PV = Extrinsic parental value. N = 451. |r|≥ .10: p < .05; |r|≥ .12: p < .01; |r|≥ .15: p < .001. Cohen's (1992) conventions for r are that .10 reflects a small effect size, .30 reflects a medium effect size, and .50 reflects a large effect size.

Table 4.3.2: Descriptive Statistics of Major Variables

Variable	M	SD
Learning strategies		
Rehearsal	3.35	1.14
Elaboration	3.91	1.17
Organization	3.52	1.20
Critical thinking	4.31	1.36
Metacognitive self-regulation	4.41	.99
Academic emotions		
Enjoyment	3.21	.70
Pride	3.34	.76
Anger	2.18	.76
Anxiety	2.54	.68
Shame	2.54	.60
Hopelessness	2.53	.85
Boredom	2.28	.86
Achievement goals		
Performance approach	4.85	1.53
Mastery avoidance	5.12	1.51
Mastery approach	5.98	1.12
Performance avoidance	4.13	1.54
Self		
Math self-concept	2.91	.70
Self-efficacy	3.24	.79
Values		
Intrinsic value	3.70	.74
Extrinsic value	3.87	.76
Environmental factors		
Lecture engagement	4.53	1.26
Harsh evaluation	4.11	.86
Evaluation focus	3.75	1.39
Parental control	2.32	.97
Peer attitude	3.14	.84
Teacher enthusiasm	3.53	1.03

Note. N = 451

Intercorrelations among the variables are in *Table 4.3.1* and the descriptive statistics of the major variables are in *Table 4.3.2*. The correlations among the scales suggest that the scales are valid measures of the motivational and cognitive constructs. Positive affect and negative affect negatively correlated with each other with positive affect and negative affect positively correlated within themselves (*rs* ranging from .24 to .84). Generally, learning strategies positively correlated with positive affect and negatively correlated with negative affect. Learning strategies subscales positively correlated with one another with rs ranging from .38 to .72. Control beliefs (self-concept and self-efficacy) positively correlated with positive affect, negatively correlated with negative affect. At the same time, control beliefs positively correlated with learning strategies use. Intrinsic value's associations with affect and learning strategies were the same as control beliefs; however, extrinsic value was positively correlated with a few negative emotions.

Students' focus on a mastery goal was found to be related with a wide range of important motivational, cognitive, and affective outcomes. Specifically, mastery goal orientation was positively related with self-efficacy, use of effective learning strategies, positive school-related affect. These findings were consistent with previous findings (Ames & Archer, 1988; L. Anderman, 1999; Kaplan & Midgley, 1999; Midgley, Anderman, & Hicks, 1995).

In general, positive environmental factors positively correlated with positive affect, learning strategy use, control beliefs, and intrinsic value. At the same time, negative environmental factors positively correlated with negative affect, negatively correlated with learning strategy use, control beliefs, and intrinsic value. Among the environmental factors, lecture engagement was positively correlated with intrinsic motivation ($r = .61$, $p < .001$), whereas harsh evaluation ($r = -.16$, $p < .001$) and evaluation focus ($r = -.34$, $p < .001$) were negatively correlated with intrinsic motivation. The findings concerning perceived classroom environment were consistent with previous findings (Church et al., 2001).

In the following subchapters intercorrelations were grouped focusing on achievement emotions, goals, learning strategies, and attributions respectively.

4.3.2 Academic emotions and associations with self, values, environmental factors, and achievement goals

- *Intercorrelations among academic emotions*

There were two categories of academic emotions, namely positive emotions of enjoyment and pride, and negative emotions of anger, anxiety, shame, hopelessness, and boredom. The two positive emotions were positively correlated with each other ($r = .72^{**}$) and negatively correlated with negative emotions except that pride and shame were not significantly correlated with each other. The negative emotions were positively correlated with each other.

- *Self and academic emotions*

According to the correlations statistics (***Table 4.3.3***), self-concept ($r = .50^{**}$, 43^{**}) and self-efficacy ($r = .55^{**}$, 45^{**}) positively correlated with enjoyment and pride; self-concept ($r = -.40^{**}$, $-.46^{**}$, $-.31^{**}$, $-.45^{**}$, $-.37^{**}$) and self-efficacy ($r = -.46^{**}$, $-.49^{**}$, $-.32^{**}$, $-.47^{**}$, $-.44^{**}$) negatively correlated with anger, anxiety, shame, hopelessness, and boredom. Therefore, the ***Hypothesis 3.1*** (*Self-concept and self-efficacy positively correlate with positive emotions; negatively correlate with negative emotions.*) was confirmed.

- *Achievement values and academic emotions*

According to the correlations statistics (***Table 4.3.3***), intrinsic value positively correlated with enjoyment ($r = .67^{**}$) and pride ($r = .51^{**}$); intrinsic value negatively correlated with anger ($r = -.49^{**}$), anxiety ($r = -.37^{**}$), shame ($r = -.13^{**}$), hopelessness ($r = -.37^{**}$), and boredom ($r = -.51^{**}$) respectively. Extrinsic value positively correlated with enjoyment ($r = $

.21**) and pride (r = .27**); Extrinsic value positively correlated with anxiety (r = .10*), shame (r = .17*), and negatively correlated with boredom (r = -.11*). Therefore, the *Hypothesis 3.2 (Achievement values correlate with both positive and negative emotions.)* was partially confirmed.

Table 4.3.3: Zero-order Correlations between Academic Emotions and Self, Values, Environmental Factors, and Achievement Goals

	Enjoyment	Pride	Anger	Anxiety	Shame	Hopelessness	Boredom
Academic emotions							
Enjoyment							
Pride	.72**						
Anger	-.68**	-.44**					
Anxiety	-.55**	-.29**	.75**				
Shame	-.24**	-.06	.47**	.65**			
Hopelessness	-.55**	-.33**	.73**	.84**	.56**		
Boredom	-.73**	-.52**	.82**	.64**	.33**	.61**	
Self							
Self-concept	.50**	.43**	-.40**	-.46**	-.31**	-.45**	-.37**
Self-efficacy	.55**	.45**	-.46**	-.49**	-.32**	-.47**	-.44**
Values							
Intrinsic value	.67**	.51**	-.49**	-.37**	-.13**	-.37**	-.51**
Extrinsic value	.21**	.27**	-.06	.10*	.17*	.07	-.11*
Environmental factors							
Lecture engagement	.72**	.48**	-.58**	-.44**	-.13**	-.45**	-.67**
Harsh evaluation	-.31**	-.20**	.32**	.25**	.17**	.23**	.33**
Evaluation focus	-.42**	-.30**	.40**	.29**	.15**	.32**	.43**
Parental control	-.24**	-.12*	.37**	.38**	.29**	.39**	.38**
Peer attitude	.47**	.42**	-.30**	-.17**	.02	-.22**	-.40**
Teacher enthusiasm	.46**	.31**	-.39**	-.26**	-.07	-.25**	-.48**
Achievement goals							
PerApproach	.17**	.29**	-.07	.12*	.17**	.10*	-.14**
MasAvoidance	.03	.12**	.10*	.30**	.27**	.21**	.04
MasApproach	.36**	.37**	-.28**	-.12**	-.02	-.17**	-.32**
PerAvoidance	-.34**	-.18**	.41**	.46**	.35**	.41**	.40**

Note. Table 4.3.3 is part of Table 4.3.1. Bivariate Pearson correlations. Sample size = 451.
*p < .05 **p< .01 (2-tailed)

- *Environmental factors and academic emotions*

In the study, lecture engagement, peer attitude, and teacher enthusiasm were defined as positive environmental factors while harsh evaluation, evaluation focus, parental control and punishment were defined as negative environmental factors. According to the correlations statistics (*Table 4.3.3*), lecture engagement (r = .72**, .48**), peer attitude (r = .47**, .42**), and teacher enthusiasm (r = .46**, .31**) positively correlated with enjoyment and pride respectively. In addition, lecture engagement, peer attitude, and teacher enthusiasm negatively correlated with anger, anxiety, shame, hopelessness, and boredom significantly except for the insignificance between peer attitude and shame, and teacher enthusiasm and shame.

It was also found that harsh evaluation (r = .32**, .25**, .17**, .23*, 33**), evaluation focus (r = .40**, .29**, .15**, .32**, .43**), and parental control (r = .37**, .38**, .29**, .39**, 38**) positively correlated with anger, anxiety, shame, hopelessness, and boredom, respectively. In addition, harsh evaluation (r = -.31**, -.20**), evaluation focus (r = -.42**, -.30**), and parental control (r = -.24**, -.12**) negatively correlated with enjoyment and pride significantly. Therefore, the *Hypothesis 3.3 (Positive environmental factors positively correlate with positive emotions; negative environmental factors positively correlate with negative emotions.)* was confirmed.

Based on the statistical results, more associations between environmental factors and academic emotions were found. Positive environmental factors positively correlated with positive emotions, negatively correlated with negative emotions; negative environmental factors positively correlated with negative emotions, and negatively correlated with positive emotions.

- *Achievement goals and academic emotions*

According to the correlation statistics (*Table 4.3.3*), performance approach goal positively correlated with enjoyment (r = .17**) and pride (r = .29**), and mastery approach goal positively correlated with enjoyment (r = .36**) and pride (r = .37**), too. Mastery avoidance goal positively correlated with anger (r = .10*), anxiety (r = .30**), shame (r = .27**), hopelessness (r = .21**); Performance avoidance goal positively correlated with anger (r = .41**), anxiety (r = .46**), shame (r = .35**), hopelessness (r = .41**), and boredom (r = .40**). Mastery avoidance and boredom were not significantly correlated. Therefore, the *Hypothesis 3.4 (Approach goals positively correlate with positive emotions; Avoidance goals positively correlate with negative emotions.) was partially confirmed.*

In addition, performance approach positively correlated with anxiety (r =.12*), shame (r = .17**), hopelessness (r = .10*), negatively correlated with boredom (r = -.14**); mastery approach negatively correlated with anger (r = -.28**), anxiety (r = -.12**), hopelessness (r = -.17**), and boredom (r = -.32**). Mastery avoidance goal positively correlated with pride (r = .12**). Performance avoidance goal negatively correlated with enjoyment (r = -.34**) and pride (r = -.18**).

Based on the above statistical results, more associations between achievement goals and academic emotions were found. The most adaptive goal was mastery approach goal, which was positively correlated with positive emotions and negatively correlated with negative emotions. The most maladaptive goal was performance avoidance goal, which was negatively correlated with positive emotions and positively correlated with negative emotions. In between were performance approach goal and mastery avoidance goal. Performance approach goal was positively correlated with positive emotions and negatively correlated with boredom but at the mean time positively correlated with anxiety, shame, and hopelessness. Mastery avoidance goal was positively correlated with pride but at the mean time positively correlated with anger, anxiety, shame, and hopelessness.

4.3.3 Achievement goals and associations with self, values, and environmental factors

- *Intercorrelations among achievement goals*

According to the correlations statistics (*Table 4.3.4*), all the four achievement goals were positively correlated with each other significantly, with Pearson r ranging from .13* to .55**. The lowest correlation existed between performance avoidance and mastery approach and the highest correlation existed between mastery approach and mastery avoidance.

- *Self and achievement goals*

According to the correlations statistics (*Table 4.3.4*), math self-concept and self-efficacy positively correlated with performance approach goal ($r = .10*$, $r = .16**$) and mastery approach goal ($r = .13**$, $r = .19**$), and negatively correlated with mastery avoidance goal ($r = -.16**$, $r = -.10*$) and performance avoidance goal ($r = -.27**$, $r = -.30**$), respectively. Therefore, the **Hypothesis 2.1** *(Self-concept and self-efficacy positively correlate with approach goals and negatively correlate with avoidance goals.)* was confirmed.

Table 4.3.4: Zero-order Correlations between Achievement Goals and Self, Values and Environmental Factors

	Performance approach	Mastery avoidance	Mastery approach	Performance avoidance
Achievement goals				
PerApproach				
MasAvoidance	.43**			
MasApproach	.48**	.55**		
PerAvoidance	.31**	.41**	.13**	
Self				
Self-concept	.10*	-.16**	.13**	-.27**
Self-efficacy	.16**	-.10*	.19**	-.30**
Values				
Intrinsic value	.23**	.09*	.37**	-.25**
Extrinsic value	.53**	.36**	.41**	.28**
Environmental factors				
Lecture engagement	.10*	.03	.34**	-.25**
Harsh evaluation	-.06	.02	-.12*	.16**
Evaluation focus	-.02	-.02	-.21**	.22**
Parental control	.07	.13**	-.16**	.30**
Peer attitude	.09	.10*	.24**	-.18**
Teacher enthusiasm	.16**	.07	.29**	-.15**

*Note. Table 4.3.4 is part of Table 4.3.1. Bivariate Pearson correlations. Sample size = 451. *$p < .05$ **$p < .01$ (2-tailed)*

- *Values and achievement goals*

According to the correlations statistics (*Table 4.3.4*), intrinsic value positively correlated with performance approach goal ($r = .23**$), mastery approach goal ($r = .37**$), and negatively correlated with performance avoidance goal ($r = -.25**$). Surprisingly, intrinsic value positively correlated with mastery avoidance goal (r = .09*). Extrinsic value positively correlated with all the four achievement goals, namely with performance approach goal (r = .53**), with mastery avoidance goal (r = .36**), with mastery approach goal (r = .41**), and performance avoidance goal (r = .28**). Therefore, the *Hypothesis 2.2 (Intrinsic value positively correlates with mastery goals; extrinsic value positively correlates with performance goals.)* was confirmed.

Based on the statistical results, intrinsic value negatively correlated with performance avoidance goal; extrinsic value positively correlated with both approach goals and avoidance goals. A tentative prediction has been made as: *Intrinsic value is more supportive than extrinsic value in math learning process.*

- Environmental factors and achievement goals

According to the correlations statistics (*Table 4.3.4*), positive environmental factors generally positively associated with approach goals: lecture engagement with performance approach (r = .10*), and with mastery approach (r = .34**); peer attitude with mastery approach (r = .24**); teacher enthusiasm with performance approach (r = .16**), and with mastery approach (r = .29**). However, the association between peer attitude and performance approach was not significant.

Negative environmental factors generally positively associated with avoidance goals: harsh evaluation with performance avoidance (r = .16**); evaluation focus with performance avoidance (r = .22**); parental control with mastery avoidance (r = .13**), and with performance avoidance (r = .30**). However, the association between harsh evaluation and mastery avoidance goal, and the association between evaluation focus and mastery avoidance goal were not significant. Therefore, the *Hypothesis 2.3 (Positive environmental factors positively associate with approach goals; negative factors positively associate with avoidance goals.)* was partially confirmed.

In addition, positive environmental factors negatively correlated with performance avoidance goal: lecture engagement with performance avoidance goal (r = -.25**); peer attitude with performance avoidance goal (r = -.18**); teacher enthusiasm with performance avoidance goal (r = -.15*). Negative environmental factors negatively correlated with mastery approach goal: harsh evaluation with mastery approach goal (r = -.12*); evaluation focus with mastery approach goal (r = -.21**); parental control with mastery approach goal (r = -.16**).

Based on the above statistics, the biggest contrast of achievement goals in terms of associations with environmental factors has emerged between mastery approach goal and performance avoidance goal. The associations between environmental factors and performance approach goal and mastery avoidance goal respectively were very few.

A tentative prediction has been made as: Environmental factors have stronger impact on mastery approach goal and performance avoidance goal than on performance approach goal and mastery avoidance goal. Some other mechanisms rather than environmental factors must have stronger impact on performance approach goal and mastery avoidance goal.

4.3.4 Learning strategies and associations with self, values, environmental factors, emotions, and goals

- *Intercorrelations among learning strategies*

According to the correlation statistics (*Table 4.3.5*), all the five learning strategies were positively correlated with one another significantly, with Pearson r ranging from .38** to .72**. The lowest correlation existed between organization and critical thinking and the highest correlation existed between elaboration and metacognitive self-regulation.

Table 4.3.5: Zero-order Correlations between Learning Strategies and Self, Values, Environmental factors, Academic Emotions, and Achievement Goals

	Rehearsal	Elaboration	Organization	Critical thinking	Metacognitive self-regulation
Learning strategies					
Rehearsal					
Elaboration	.59**				
Organization	.62**	.59**			
Critical thinking	.40**	.60**	.38**		
Metacognitive self-regulation	.49**	.72**	.52**	.68**	
Self					
Math self-concept	.16**	.37**	.24**	.46**	.48**
Self-efficacy	.18**	.41**	.25**	.48**	.51**
Values					
Intrinsic value	.25**	.40**	.36**	.41**	.50**
Extrinsic value	.14**	.10*	.14**	.16**	.14**
Environmental factors					
Lecture engagement	.27**	.45**	.32**	.43**	.54**
Harsh evaluation	.01	-.07	-.05	-.05	-.11*
Evaluation focus	-.12*	-.25**	-.18**	-.20**	-.29**
Parental control	-.01	-.12*	-.14**	-.20**	-.31**
Peer attitude	.20**	.26**	.25**	.21**	.30**
Teacher enthusiasm	.15**	.27**	.16**	.29**	.36**
Academic emotions					
Enjoyment	.37**	.56**	.43**	.50**	.64**
Pride	.37**	.48**	.39**	.44**	.51**
Anger	-.21**	-.43**	-.31**	-.44**	-.58**
Anxiety	-.11*	-.35**	-.18**	-.41**	-.49**
Shame	-.002	-.14**	-.03	-.21**	-.26**
Hopelessness	-.14**	-.36**	-.22**	-.38**	-.50**
Boredom	-.29**	-.44**	-.35**	-.44**	-.62**
Achievement goals					
PerApproach	.15**	.15**	.19**	.23**	.20**
MasAvoidance	.21**	.05	.12**	.05	.06
MasApproach	.26**	.29**	.28**	.33**	.44**
PerAvoidance	.05	-.15**	-.02	-.22**	-.27**

Note. Table 4.3.5 is part of Table 4.3.1. Bivariate Pearson correlations. N = 451. *p< .05 **p< .01 (2-tailed).

- *Self and learning strategies*

According to the correlation statistics (*Table 4.3.5*), math self-concept and self-efficacy positively correlated with rehearsal (r = .16**, r = .18**), elaboration (r = .37**, r = .41**), organization (r = .24**, r = .25**), critical thinking (r = .46**, r = .48**), and metacognitive self-regulation (r = .48**, r = .51**), respectively. Therefore, the **Hypothesis 5.1** *(Math self-concept and self-efficacy in learning mathematics positively correlate with learning strategies use in mathematics learning process.)* was confirmed.

- *Values and learning strategies*

According to the correlation statistics (*Table 4.3.5*), intrinsic value and extrinsic value positively correlated with rehearsal (r = .25**, r = .14**), elaboration (r = .40**, r = .10*), organization (r = .36**, r = .14**), critical thinking (r = .41**, r = .16**), and metacognitive self-regulation (r = .50**, r = .14**), respectively. Therefore, the **Hypothesis 5.2** *(Achievement values positively correlate with strategy use)* was confirmed.

- *Environmental factors and learning strategies*

Lecture engagement, peer attitude, and teacher enthusiasm were defined as positive environmental factors; harsh evaluation, evaluation focus, and parental control and punishment were defined as negative environmental factors in the study. According to the correlation statistics (*Table 4.3.5*), lecture engagement, peer attitude, and teacher enthusiasm positively correlated with rehearsal (r = .27**, r = .20**, r = .15**), elaboration (r = .45**, r = .26**, r = .27**), organization (r = .32** r = .25**, r = .16**), critical thinking (r = .43**, r = .21**, r = .29**), and metacognitive self-regulation (r = .54**, r = .30**, r = .36**), respectively.

The associations between negative environmental factors and learning strategies were more complex than the associations between positive environmental factors and learning strategies. Harsh evaluation only negatively correlated with metacognitive self-regulation (r = -.11*). Evaluation focus negatively correlated with rehearsal (r = -.12*), elaboration (r = -.25**), organization (r = -.18**), critical thinking (r = -.20**), and metacognitive self-regulation (r = -.29**). Parental control negatively correlated with elaboration (r = -.12*), organization (r = -.14**), critical thinking (r = -.20**), and metacognitive self-regulation (r = -.31**).

Based on the above statistics, rather clear associations between environmental factors and learning strategies have emerged. Positive environmental factors positively associate with learning strategies; negative environmental factors generally negatively associate with learning strategies. However, the strength of environmental factors in predicting learning strategies use in learning mathematics must be considered in conjunction with other variables, e.g., self-concept, academic emotions, and achievement goals, etc..

- *Academic emotions and learning strategies*

According to the correlation statistics (*Table 4.3.5*), positive affect positively correlated with learning strategy use. Enjoyment and pride positively correlated with rehearsal (r = .37**, r = .37**), elaboration (r = .56**, r = .48**), organization (r = .43**, r = .39**), critical thinking (r = .50**, r = .44**), and metacognitive self-regulation (r = .64**, r = .51**), respectively.

Anger, anxiety, hopelessness, and boredom negatively correlated with rehearsal (r = -.21**, r = -.11*, r = -.14**, r = -.29**), elaboration (r = -.43**, r = -.35**, r = -.36**, r = -.44**), organization (r = -.31**, r = -.18**, r = -.22**, r = -.35**), critical thinking (r = -.44**, r = -.41**, r = -.38**, r = -.44**), and metacognitive self-regulation (r = -.58**, r = -.49**, r = -.50**, r = -.62**), respectively. Shame negatively correlated with elaboration (r = -.14**), critical thinking (r = -.21**), and metacognitive self-regulation (r = -.26**). However, shame was not correlated with rehearsal and organization significantly. Therefore, the *Hypothesis 5.3 (Positive emotions positively correlate with learning strategy use; negative emotions negatively correlate with learning strategy use)* was partially confirmed.

Achievement goals and learning strategies

According to the correlation statistics (*Table 4.3.5*), performance approach goal and mastery approach goal positively correlated with the five learning strategies, namely rehearsal (r = .15**, r = .26**), elaboration (r = .15**, r = .29**), organization (r = .19**, r =.28**), critical thinking (r = .23**, r = .33**), and metacognitive self-regulation (r = .20**, r =.44**), respectively.

Mastery avoidance goal positively correlated with rehearsal (r = .21**) and organization (r = .12**). Performance avoidance goal negatively correlated with elaboration (r = -.15**), critical thinking (r = -.22**), and metacognitive self-regulation (r = -.27**). The opposite valences of the correlations between the two avoidance goals and learning strategy use made the association between avoidance goals and learning strategy use insignificant (r = -.02).

Therefore, a clear pattern concerning achievement goals and learning strategies has emerged: Approach goals positively associated with applying learning strategies in learning math; Avoidance goal (combined out of mastery avoidance and performance avoidance) did not associate with applying learning strategies. It could be expected that avoidance goal would not function as a mediator on learning strategy use.

4.3.5 Attributions and academic emotions

Ranked causal attributions, and internal and external attributions for success and failure

Table (*Table 4.3.6*) shows the means for internal and external attributions as well as specific causes ranked in order of importance in success and failure situations. Effort and learning strategies were the most important causes for both success and failure. Luck and task difficulty were the least important causes for success. Luck and ability were the least important causes for failure. In line with Salili's (1995) findings, effort ranked top one among the causes for success and failure in collectivist culture. The potential role of ability in explaining success was very much downplayed; this finding was in line with Shikanai's (1978) study of college students. Indeed, in the current study ability was perceived to be more important after a success than after a failure, whereas task difficulty (ease) was perceived to be more important after a failure than after a success.

Table 4.3.6: Means for Causal Attributions in Ranked Order

Rank	Success			Failure		
	Cause	Mean	S.D.	Cause	Mean	S.D.
1	Effort	3.96	(0.93)	Effort	3.91	(1.06)
2	Learning strategies	3.71	(0.93)	Learning strategies	3.81	(0.93)
3	Teaching quality	3.52	(1.03)	Task difficulty	2.81	(1.08)
4	Ability	3.01	(1.08)	Teaching quality	2.52	(1.13)
5	Luck	2.79	(1.15)	Luck	2.32	(1.05)
6	Task difficulty	2.71	(1.06)	Ability	2.31	(1.10)
Internal	cause	10.68	(2.09)		10.03	(2.09)
External	cause	9.02	(2.22)		7.64	(2.32)

Table 4.3.7: Mean Differences and Standard Deviations of Internal Attribution and External Attribution for Success and Failure

	Mean difference	S.D.	T	Df
Pair 1 Internal causes for success – External causes for success	1.65	2.71	13.07***	460
Pair 2 Internal causes for failure – External causes for failure	2.39	2.56	20.10***	460

Note. ****p < .001 (2-tailed).*

Paired sample t test showed that on the whole subjects made more internal than external attributions for both success and failure (t = 13.07, and t = 20.10, p < .001 respectively). The interaction effect showed that there were higher internal attributions for success than for failure (t = 4.96, p < .001) and there were higher external attributions for success than for failure (t = 10.11, p < .001).

These results supported that Chinese high school students attributed both success and failure more to internal than external causes. A strong cultural impact on attributional style is apparent. Ability was not perceived as important as effort, learning strategies, and teaching quality, etc... Effort was regarded as the most important cause for success and failure. This pattern confirms earlier findings about Chinese attributional styles (e.g. Fry & Ghosh, 1980; Salili, 1995; Salili, Hwang & Choi, 1989). Teaching quality was also an important cause for success and failure outcome according to Chinese high school students' point of view, which reflected another Chinese belief, "Good students come from good teachers" (Salili, 1995).

Attributions and achievement emotions

- *Attributions and enjoyment*

Six potential predictors, namely ability, effort, luck, task ease, learning strategies, and quality of instruction, were put into the model for predicting enjoyment. Forward regression analysis was applied to identify the best model for predicting enjoyment.

Table 4.3.8: Predictors for Positive Emotions with Attributions as Independent Variables

		β	t	R^2 adjusted	$F_{(4, 452)}$
ENJOYMENT AS	Teaching quality	.32	7.46***	.19	27.47***
DEPENDENT	Ability	.19	4.49***		
VARIABLE	Luck	-.16	-3.61***		
	Effort	.13	2.96**		

		β	t	R^2 adjusted	$F_{(3, 450)}$
PRIDE AS	Teaching quality	.23	5.34***	.17	31.06***
DEPENDENT	Ability	.21	4.87***		
VARIABLE	Effort	.20	4.49***		

Note. ***p < .01.* ****p < .001.*

Model summary (*Table 4.3.8*) showed that the best model explained 19% (adjusted R square) variances for enjoyment. MANOVA statistics showed that Model 4 was significant (F (4, 452) = 27.47 at .001 level). The significant predictors included quality of instruction (ß = .32, t = 7.46 at .001 level), ability (ß = .19, t = 4.49 at .001 level), luck (ß = -.16, t = -3.61, at .001 level), and effort (ß = .13, t = 2.96 at .01 level). Based on the above statistics, two external causes, quality of instruction and luck (negative predictor), and two internal causes, ability and effort, were identified as antecedents for enjoyment.

- *Attributions and pride*

Six potential predictors, namely ability, effort, luck, task ease, learning strategies, and quality of instruction, were put into the model for predicting pride. Forward regression analysis was applied to identify the best model for predicting pride.

Model summary (*Table 4.3.8*) showed that the best model explained 17% (adjusted R square) variances for pride. MANOVA statistics showed that the model was significant (F (3, 450) = 31.06 at .001 level). The significant predictors included quality of instruction (ß = .23, t = 5.34 at .001 level), ability (ß = .21, t = 4.87 at .001 level), and effort (ß = .20, t = 4.49 at .001 level).

Therefore, the ***Hypothesis 4.1*** *(Attributional antecedents for pride are internal causes, e.g., ability and/or effort.)* was confirmed. Based on the above statistical results, one external cause: quality of instruction and two internal causes were found to be main antecedents for positive emotions.

- *Attributions and anger*

Six potential predictors, namely lack of ability, effort, luck, task difficulty, poor learning strategies, and poor quality of instruction, were put into the model for predicting anger. Forward regression analysis was applied to identify the best model for predicting anger.

Model summary (***Table 4.3.9***) showed that the best model explained 16% (adjusted R square) variances for anger. MANOVA statistics showed that the model was significant (F (3, 452) = 28.82 at .001 level). The significant predictors were lack of ability (ß = .20, t = 4.04 at .001 level), poor quality of instruction (ß = .21, t = 4.79 at .001 level), and task difficulty (ß = .15, t = 3.03 at .01 level).

Therefore, the ***Hypothesis 4.2*** *(Attributional antecedents for anger are controllable causes, e.g., effort, learning strategy use, or quality of instruction.)* was confirmed. However, effort and learning strategies use were not antecedents for anger. Based on the above statistics, two uncontrollable causes, namely ability and task difficulty were also antecedents for anger.

- *Attributions and anxiety*

Six potential predictors, namely lack of ability, effort, luck, learning strategies, task difficulty, and poor quality of instruction were put into the model for predicting anxiety. Forward regression analysis was applied to identify the best model for predicting anxiety.

Model summary (***Table 4.3.9***) showed that the best model 3 explained 19% (adjusted R square) variances for anxiety. MANOVA statistics showed that Model 3 was significant (F (3, 452) = 36.18 at .001 level). The significant predictors were lack of ability (ß = .27, t = 5.54 at .001 level), task difficulty (ß = .18, t = 3.69 at .001 level), and poor quality of instruction (ß = .15, t = 3.39 at .001 level).

Based on the above statistics, one internal uncontrollable cause, ability, two external causes, task difficulty and quality of instruction were identified as antecedents for anxiety.

- *Attributions and shame*

Six potential predictors, namely lack of ability, effort, luck, task difficulty, poor learning strategies, and poor quality of instruction, were put into the model for predicting shame. Forward regression analysis was applied to identify the best model for predicting shame.

Model summary (***Table 4.3.9***) showed that the best model 3 explained 11% (adjusted R square) variances for shame. MANOVA statistics showed that model 3 was significant (F (3, 450) = 19.99 at .001 level). The significant predictors were lack of ability (ß = .23, t = 5.16 at .001 level), poor quality of instruction (ß = .17, t = 3.69 at .001 level), and poor learning strategies (ß = .12, t = 2.63 at .01 level).

Therefore, the ***Hypothesis 4.3*** *(Attributional antecedents for shame are internal and uncontrollable causes, e.g., lack of ability.)* was confirmed. Based on the above statistics, one internal and controllable cause, poor learning strategy use, and one external cause, poor quality of instruction were also antecedents for shame.

Table 4.3.9: Predictors for Negative Emotions with Attributions as Independent Variables

		β	t	R^2 adjusted	F (3, 452)
ANGER AS DEPENDENT VARIABLE	Lack of ability Poor teaching quality Task difficulty	.20 .21 .15	4.04*** 4.79*** 3.03**	.16	28.82***
		β	t	R^2 adjusted	F (3, 452)
ANXIETY AS DEPENDENT VARIABLE	Lack of ability Task difficulty Poor teaching quality	.27 .18 .15	5.54*** 3.69*** 3.39***	.19	36.18***
		β	t	R^2 adjusted	F (3, 452)
SHAME AS DEPENDENT VARIABLE	Lack of ability Poor teaching quality Poor learning strategies	.23 .17 .12	5.16*** 3.69*** 2.63**	.11	19.99***
		β	t	R^2 adjusted	F (3, 450)
HOPELESSNESS AS DEPENDENT VARIABLE	Lack of ability Poor teaching quality Task difficulty	.24 .16 .16	4.93*** 3.60*** 3.35***	.17	31.01***
		β	t	R^2 adjusted	F (3, 453)
BOREDOM AS DEPENDENT VARIABLE	Poor teaching quality Task difficulty Lack of ability	.28 .16 .15	6.49*** 3.36*** 3.01**	.18	33.29***

Note. **$p < .01$. ***$p < .001$.

- *Attributions and hopelessness*

Six potential predictors, namely lack of ability, effort, luck, task difficulty, poor learning strategies, and poor quality of instruction, were put into the model for predicting hopelessness. Forward regression analysis was applied to identify the best model for predicting hopelessness. Model summary (*Table 4.3.9*) showed that the best model explained 17% (adjusted R square) variances for hopelessness. MANOVA statistics showed that model 3 was significant (F (3, 450) = 31.01 at .001 level). The significant predictors were lack of ability (ß = .24, t = 4.93 at .001 level), poor quality of instruction (ß = .16, t = 3.60 at .001 level), and task difficulty (ß = .16, t = 3.35 at .001 level).

Therefore, the *Hypothesis 4.4 (Attributional antecedents for hopelessness are stable internal causes, e.g., lack of ability.)* was confirmed. Based on the above statistics, two external causes, poor quality of instruction and task difficulty were antecedents for hopelessness.

- *Attributions and boredom*

Six potential predictors, namely lack of ability, effort, luck, task difficulty, poor learning strategies, and poor quality of instruction, were put into the model for predicting boredom. Forward regression analysis was applied to identify the best model for predicting boredom. Model summary (*Table 4.3.9*) showed that the best model 3 explained 18% (adjusted R square) variances for boredom. MANOVA statistics showed that Model 3 was significant (F (3, 453) = 33.29 at .001 level). The significant predictors were poor quality of instruction (ß = .28, t = 6.49 at .001 level), task difficulty (ß = .16, t = 3.36 at .001 level), and lack of ability (ß = .15, t = 3.01 at .01 level).

Based on the above statistics, antecedents for boredom were two external causes, poor quality of instruction, and task difficulty, and one internal uncontrollable cause, lack of ability.

- *Summary: attributions and academic emotions*

Based on the above regression statistics, a clear pattern of antecedents for academic emotions has emerged. Ability and teaching quality were antecedents for all the seven emotions, which highlighted the importance of internal uncontrollable antecedent of ability, and external antecedent of teaching quality. Effort predicted only positive emotions but not negative emotions. Task difficulty only predicted negative emotions but not positive emotions. Luck negatively predicted enjoyment and inappropriate learning strategies predicted shame. Since the ability of a learner can not be enhanced quickly in a short time span, it is meaningful to pay attention on improving quality of instruction. In this way, it is possible to shift learners' emotion experiences in learning mathematics toward more positive dimension.

Although the Chinese high school sample students experienced negative affect frequently in learning mathematics and (lack of) ability was one of the most frequent predictors for negative emotions, they attributed their success and failure more to effort than to ability (*Table 4.3.6*), which (effort) is an internal and controllable antecedent and consequently facilitative for improving one's performance in mathematics. It reflects more or less the adaptive attributional style of adolescents in a collectivist culture.

That which is the stronger predictor for math GPA, attribution or emotion, is an emerged question here. It is also interesting to find out whether attribution and emotion co-predict math GPA. Therefore, forward regression analysis was applied to test the predictor(s) for math GPA. Emotions of enjoyment, pride, anger, anxiety, shame, hopelessness and boredom, six attributions for success, and six attributions for failure were loaded on the model for predicting math GPA.

Model summary showed that the best model 2 explained 33% (adjusted R square) variances for math GPA. MANOVA statistics showed that Model 2 was significant (F (2, 354) = 89.32 at .001 level). The significant predictors were boredom (ß = -.43, t = -7.79 at .001 level), and hopelessness (ß = -.21, t = -3.81 at .001 level) (*Table 4.3.10*). These results were in line with Weiner's (1985) attributional view of the emotion processes, which states that following outcome appraisal and the immediate affective reaction, a causal ascription will be sought and a different set of emotions is then generated by the chosen attribution(s). In the mathematics learning context, attributional thinking was one of the intermediate steps towards the formation of academic emotions, which (boredom and hopelessness) were finally found to predict mathematics achievement. The common predictors for boredom and hopelessness were poor teaching quality, task difficulty, and lack of ability (*Table 4.3.9*). In another word, boredom and hopelessness would be frequently experienced by low achievers; Boredom and hopelessness would often associate with an adolescent's evaluation of his mathematical learning ability as being low, task difficulty as being high, and teacher's instruction quality as being poor.

Table 4.3.10: Predictors for Math GPA with Emotions and Attributions as Independent Variables

	β	t	R^2 adjusted	F (2, 354)
Boredom	-.43	-7.79***	.33	89.32***
Hopelessness	-.21	-3.81***		

Note. ****p < .001*

Furthermore, none of the attributional antecedents for success or failure was found to predict math GPA. The association between emotions and achievement was found to be more immediate and stronger than the association between attributions and achievement. In another word, although learners in a collectivist culture have been trained to adopt and get used to effort-leading-to-achievement attribution thinking style, their frequent emotional experiences saliently predict their achievement rather than their trained thinking style. A split between feelings and thinking has displayed itself in the context of mathematics learning in a collectivist culture of China. A vivid example could be: a Chinese adolescent who has been trained to adopt the effort-leading-to-achievement thinking module, feels bored with learning mathematics but still tries to persuade himself to invest more effort on his math subject. Sadly his feelings have more say on his achievement than his internalized achievement attribution thinking style.

The following questions need to be clarified based on cross-cultural studies in the future. How does western adolescent learners' achievement attribution thinking style look like and associate with their feelings on their achievement? Do adolescents' emotions in math learning vary with their attribution thinking styles? How much weight does effort oriented attribution load on achievement compared with ability oriented attribution? The above questions need to be clarified before we make the intuitive statement that "effort-leading-to-achievement attribution is more facilitative to achievement than ability-leading-to-achievement attribution".

4.4 Mediator effect

4.4.1 Math self-concept as a mediator

Prerequisites for mediator testing

It was the researcher's interest to test whether math self-concept could mediate environmental factors' impact on emotions. According to the frequently cited paper on the moderator-mediator distinction written by Baron and Kenny (1986), several requirements must be met before a mediating relationship can be discussed and analysized (see also Kenny, Kashy & Bolger, 1998). The predominant relationship is labeled "c," and is the path from the independent to the dependent variable. The mediating path has two parts, comprised of "a," the path connecting the independent variable to the potential mediator, and "b," the path connecting the mediator to the dependent variable.

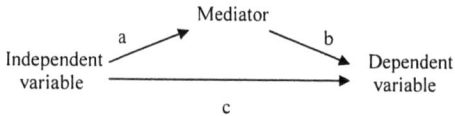

Note: source Baron and Kenny (1986)

Baron and Kenny (1986) stated that to claim a mediating relationship, there should be a significant relationship between the independent variable and the mediator. The next step is to show that there is a significant relationship between the mediator and the dependent variable. A significant relationship between the independent variable and the dependent variable should exist simultaneously. The three prerequisites for mediator testing are the three significant paths among the independent, potential mediator, and the dependent variable. The next important step is to use the mediator and the independent variable to predict the dependent variable simultaneously. If the previously significant path between the independent and dependent variables is now to a large extend reduced. The reduced significant path can be insignificant, while the most likely occurrence is that Path c transforms into a weaker, though perhaps still significant, path.

Self-concept between positive environmental factors and positive emotions

Teacher enthusiasm, peer attitude, and lecture engagement were defined as positive environmental factors in the study. The variable of positive emotions was the mean of pride, and enjoyment.

- *Math self-concept as a potential mediator between teacher enthusiasm and positive emotions*

From the correlation table, the three paths between the independent variable of teacher enthusiasm and the mediator of math self-concept, between the mediator of math self-concept and the dependent variable of positive emotion, and between the independent variable of teacher enthusiasm and the dependent variable of positive emotion were all significantly correlated (Pearson r = .20, r = .50, r = .42, p < 0.01).

Table 4.4.1: Pearson Correlations among Teacher Enthusiasm, Math Self-concept and Positive Emotions

	Teacher enthusiasm	Math self-concept	Positive emotions
Teacher enthusiasm	1.00	.20**	.42**
Math self-concept	.20**	1.00	.50**
Positive emotions	.42**	.50**	1.00

Note. **Correlation is significant at the 0.01 level (2-tailed).

Table 4.4.2: Coefficients of Teacher Enthusiasm and Math Self-concept

Model		B	SE B	β	T	Correlations Zero-order	Semi-Partial
1	(Constant)	2.32	.10		22.65***		
	Teacher enthusiasm	.27	.03	.42	9.75***	.42	.42
2	(Constant)	1.28	.13		10.02***		
	Teacher enthusiasm	.22	.03	.33	8.59***	.42	.32
	Math self-concept	.42	.04	.44	11.39***	.50	.43

Note. Dependent variable: Positive emotions. ***p < .001.

After the three prerequisites were satisfied, the independent variable of teacher enthusiasm alone and the independent variable of teacher enthusiasm and the mediator of math self-concept were used to predict the dependent variable of positive emotions respectively. From the coefficient tables, it is observable that the direct path between teacher enthusiasm and positive emotion became less important (zero-order correlation 0.42 decreased to 0.32 due to the mediating effect of math self-concept) but it was still significant. Sobel provided a good solution to this complex situation, in which the direct path remains significant, though at a lower value. The method to test for a mediating relationship is to ask if the complete mediating path from independent variable to mediator to dependent variable is significant (Sobel, 1982).

The regression coefficients and their standard errors for the two paths in the mediating chain were shown in *Table 4.4.3*.

Table 4.4.3: Regression Coefficients and Standard Errors for Two Paths of Mediating Path from Teacher Enthusiasm, Math self-concept to Positive Emotions

Path a		Path b	
Teacher enthusiasm → Math self-concept		Math self-concept → Positive emotions	
$ß_a$	0.196	$ß_b$	0.504
s_a	0.031	s_b	0.039
t	4.274***	t	12.424***

Note. ***$p < .001$

We know $t = ß/S_ß$. The regression coefficient for the path from Teacher enthusiasm to Self-concept to Positive emotions is equal to $ß_a \times ß_b$, where a and b refer to the relevant paths. Path c is the direct path from teacher enthusiasm to positive emotions.

$$S_{ß_a ß_b} = \sqrt{ß_a^2 s_b^2 + ß_b^2 s_a^2 - s_a^2 s_b^2}$$

$$= \sqrt{0.196^2 (0.039)^2 + 0.504^2 (0.031)^2 - (0.031)^2 (0.039)^2}$$

$$= 0.017$$

The mediated path coefficient was 0.196 x 0.504 and its standard error (0.017), therefore the t ratio was calculated as

$$t = \frac{ß_a ß_b}{S_{ß_a ß_b}} = \frac{0.196 \times 0.504}{0.017} = 5.81$$

Sobel (1982) stated that this ratio is asymptotically normally distributed, which, for large samples, would lead to rejection of the null hypothesis at $a = 0.05$ when the ratio exceeds ± 1.96. In this case the mediated path was clearly significant. Therefore it can be concluded that convincing evidence of a strong mediating pathway from teacher enthusiasm through math self-concept to positive emotions existed.

- *Math self-concept as a potential mediator between peer attitude and positive emotions*

Table 4.4.4: Pearson Correlations among Peer Attitude, Math Self-concept, and Positive Emotions

	Peer attitude	Math self-concept	Positive emotions
Peer attitude	1.00	.22**	.48**
Math self-concept	.22**	1.00	.50**
Positive emotions	.48**	.50**	1.00

Note. **Correlation is significant at the 0.01 level (2-tailed).

From the correlation table, the three paths between the independent variable of peer attitude and the mediator of math self-concept, between the mediator of math self-concept and the dependent variable of positive emotions, and between the independent variable of peer attitude and the dependent variable of positive emotions were all significantly correlated (Pearson r = .22, r = .50, r = .48, p < 0.01).

Table 4.4.5: Coefficients of Peer Attitude and Math Self-concept

Model		B	SE B	β	T	Correlations Zero-order	Semi-Partial
1	(Constant)	2.07	.11		19.06***		
	Peer attitude	.39	.03	.48	11.54***	.48	.48
2	(Constant)	1.12	.13		8.70***		
	Peer attitude	.31	.03	.39	10.28***	.48	.38
	Math self-concept	.41	.04	.42	11.18***	.50	.41

Note. Dependent variable: Positive emotions. ****p < .001.*

Table 4.4.6: Regression Coefficients and Standard Errors for Two Paths of Mediating Path from Peer Attitude, Math Self-concept, to Positive Emotions

Path a		Path b	
Peer attitude ⟶ Math self-concept		Math self-concept ⟶ Positive emotions	
$β_a$	0.218	$β_b$	0.504
s_a	0.038	s_b	0.039
t	4.762***	t	12.424***

Note. ****p < .001*

After the prerequisites for mediator testing were satisfied, the independent variable of peer attitude alone and the independent variable of peer attitude together with the mediator of math self-concept were used to predict the dependent variable of positive emotions. From the coefficient tables, it is observable that the direct path between peer attitude and positive emotions became less important (zero-order correlation 0.48 decreased to 0.38 due to the mediating effect of math self-concept) but it was still significant.

Based on Sobel's (1982) solution for this complex situation, in which the direct path remains significant, though at a lower value, the method to test for a mediating relationship is to ask if the complete mediating path from independent variable to mediator to dependent variable is significant. The regression coefficients and their standard errors for the two paths in the mediating chain were shown in *Table 4.4.6.*

We know $t = ß/S_ß$. The regression coefficient for the path from Peer attitude to Self-concept to Positive emotions is equal to $ß_a \times ß_b$, where a and b refer to the relevant paths. Path c is the direct path from peer attitude to positive emotions.

$$S_{ß_a ß_b} = \sqrt{ß_a^2 s_b^2 + ß_b^2 s_a^2 - s_a^2 s_b^2}$$

$$= \sqrt{0.218^2 (0.039)^2 + 0.504^2 (0.038)^2 - (0.038)^2 (0.039)^2}$$

$$= 0.021$$

The mediated path coefficient was 0.218 x 0.504 and its standard error (0.021), therefore the t ratio was calculated as

$$t = \frac{ß_a ß_b}{S_{ß_a ß_b}} = \frac{0.218 \times 0.504}{0.021} = 5.232$$

Sobel (1982) stated that this ratio is asymptotically normally distributed, which, for large samples, would lead to rejection of the null hypothesis at a = 0.05 when the ratio exceeds ±1.96. In this case the mediated path was clearly significant. Therefore it can be concluded that convincing evidence of a strong mediating pathway from peer attitude through math self-concept to positive emotions existed.

- *Math self-concept as a potential mediator between lecture engagement and positive emotions*

From the correlation table (***Table 4.4.7***), the three paths between the independent variable of lecture engagement and the mediator of math self-concept, between the mediator of math self-concept and the dependent variable of positive emotions, and between the independent variable of lecture engagement and the dependent variable of positive emotions were all significantly correlated (Pearson r = .39, r = .50, r = .65, p < 0.01).

Table 4.4.7: Pearson Correlations among Lecture Engagement, Math Self-concept and Positive Emotions

	Lecture engagement	Math self-concept	Positive emotions
Lecture engagement	1.00	.39**	.65**
Math self-concept	.39**	1.00	.50**
Positive emotions	.65**	.50**	1.00

*Note. **Correlation is significant at the 0.01 level (2-tailed).*

After the prerequisites for mediator testing were satisfied, the independent variable of lecture engagement alone and the independent variable of lecture engagement together with the mediator of math self-concept were used to predict the dependent variable of positive emotions.

From the coefficient table (*Table 4.4.8*), it is observable that the direct path between lecture engagement and positive emotions became less important (zero-order correlation 0.65 decreased to 0.49 due to the mediating effect of math self-concept) but it was still significant.

Table 4.4.8: Coefficients of Lecture Engagement and Math self-concept

Model	B	SE B	β	T	Zero-order	Semi-Partial
1 (Constant)	1.72	.09		19.24***		
Lecture engagement	.34	.02	.65	18.07***	.65	.65
2 (Constant)	1.17	.11		10.84***		
Lecture engagement	.28	.02	.53	14.58***	.65	.49
Math self-concept	.29	.04	.30	8.08***	.50	.27

Note. Dependent variable: Positive emotions. $***p < .001$.

Table 4.4.9: Regression Coefficients and Standard Errors for Two Paths of Mediating Path from Lecture Engagement, Math Self-concept, to Positive Emotions

Path a		Path b	
Lecture engagement → Math self-concept		Math self-concept → Positive emotions	
$β_a$	0.389	$β_b$	0.504
s_a	0.024	s_b	0.039
t	9.044***	t	12.424***

Note. $***p < .001$

Based on Sobel's (1982) solution for this complex situation, in which the direct path remains significant, though at a lower value, the method to test for a mediating relationship is to ask if the complete mediating path from independent variable to mediator to dependent variable is significant. The regression coefficients and their standard errors for the two paths in the mediating chain were shown in *Table 4.4.9*.

We know $t = β/S_β$. The regression coefficient for the path from Lecture engagement to Math self-concept to Positive emotions is equal to $β_a \times β_b$, where a and b refer to the relevant paths. Path c is the direct path from lecture engagement to positive emotion.

$$S_{β_aβ_b} = \sqrt{β_a^2 s_b^2 + β_b^2 s_a^2 - s_a^2 s_b^2}$$
$$= \sqrt{0.389^2 (0.039)^2 + 0.504^2 (0.024)^2 - (0.024)^2 (0.039)^2}$$
$$= 0.019$$

The mediated path coefficient was 0.389 x 0.504 and its standard error (0.019), therefore the t ratio was calculated as

$$t = \frac{\beta_a \beta_b}{S_{\beta_a \beta_b}} = \frac{0.389 \times 0.504}{0.019} = 10.32$$

Sobel (1982) stated that this ratio is asymptotically normally distributed, which, for large samples, would lead to rejection of the null hypothesis at a = 0.05 when the ratio exceeds ± 1.96. In this case the mediated path was clearly significant. Therefore it can be concluded that convincing evidence of a strong mediating pathway from lecture engagement through math self-concept to positive emotions existed.

Self-concept between negative environmental factors and negative emotions

Harsh evaluation, evaluation focus, and parental control were defined as negative environmental factors in the current study. The variable of negative emotions was the mean of anger, anxiety, hopelessness, shame, and boredom.

- *Math self-concept as a potential mediator between harsh evaluation and negative emotions*

Table 4.4.10: Pearson Correlations among Harsh Evaluation, Math Self-concept, and Negative Emotions

	Harsh evaluation	Math self-concept	Negative emotions
Harsh evaluation	1.00	-.20**	.30**
Math self-concept	-.20**	1.00	-.47**
Negative emotions	.30**	-.47**	1.00

Note. **Correlation is significant at the 0.01 level (2-tailed).

From the correlation table (*Table 4.4.10*), the three paths between the independent variable of harsh evaluation and the mediator of math self-concept, between the mediator of math self-concept and the dependent variable of negative emotions, and between the independent variable of harsh evaluation and the dependent variable of negative emotions were all significantly correlated (Pearson r = -.20, r = -.47, r = .30, p < 0.01).

After the prerequisites for mediator testing were satisfied, the independent variable of harsh evaluation alone and the independent variable of harsh evaluation together with the mediator of math self-concept were used to predict the dependent variable of negative emotions. From the coefficient tables, it is observable that the direct path between harsh evaluation and negative emotions became less important (zero-order correlation 0.30 decreased to 0.21 due to the mediating effect of math self-concept) but it was still significant.

Table 4.4.11: Coefficients of Harsh Evaluation and Math Self-concept

Model	B	SE B	β	T	Correlations Zero-order	Semi-Partial
1 (Constant)	1.49	.14		10.73***		
Harsh evaluation	.22	.03	.30	6.77***	.30	.30
2 (Constant)	2.90	.18		15.78***		
Harsh evaluation	.16	.03	.22	5.24***	.30	.21
Math self-concept	-.39	.04	-.43	-10.45***	-.47	-.42

Note. Dependent variable: Negative emotions. ****p < .001.*

Based on Sobel's (1982) solution for this complex situation, in which the direct path remains significant, though at a lower value, the method to test for a mediating relationship is to ask if the complete mediating path from independent variable to mediator to dependent variable is significant. The regression coefficients and their standard errors for the two paths in the mediating chain were shown in *Table 4.4.12*.

Table 4.4.12: Regression Coefficients and Standard Errors for Two Paths of Mediating Path from Harsh Evaluation, Math Self-concept, to Negative Emotions

Path a		Path b	
Harsh evaluation ⟶ Math self-concept		Math self-concept ⟶ Negative emotions	
$ß_a$	-0.200	$ß_b$	-0.473
s_a	0.037	s_b	0.038
t	-4.382***	t	-11.442***

Note. ****p < .001*

We know $t = ß/S_ß$. The regression coefficient for the path from Harsh evaluation to Math self-concept to Negative emotions is equal to $ß_a \times ß_b$, where a and b refer to the relevant paths. Path c is the direct path from harsh evaluation to negative emotions.

$$S_{ß_a ß_b} = \sqrt{ß_a^2 s_b^2 + ß_b^2 s_a^2 - s_a^2 s_b^2}$$

$$= \sqrt{(-0.200)^2 (0.038)^2 + (-0.473)^2 (0.037)^2 - (0.037)^2 (0.038)^2}$$

$$= 0.019$$

The mediated path coefficient was (-0.200) x (-0.473) and its standard error (0.019), therefore the t ratio was calculated as

$$t = \frac{ß_a ß_b}{S_{ß_a ß_b}} = \frac{(-0.200) \times (-0.473)}{0.019} = 4.98$$

Sobel (1982) stated that this ratio is asymptotically normally distributed, which, for large samples, would lead to rejection of the null hypothesis at a = 0.05 when the ratio exceeds ±1.96. In this case the mediated path was clearly significant. Therefore it can be concluded that convincing evidence of a strong mediating pathway from harsh evaluation through math self-concept to negative emotions existed.

- *Math self-concept as a potential mediator between evaluation focus and negative emotions*

From the correlation table, the three paths between the independent variable of evaluation focus and the mediator of math self-concept, between the mediator of math self-concept and the dependent variable of negative emotions, and between the independent variable of evaluation focus and the dependent variable of negative emotions were all significantly correlated (Pearson r = -.20, r = -.47, r = .38, p < 0.01).

Table 4.4.13: Correlations among Evaluation focus, Math Self-concept, and Negative Emotions

	Evaluation focus	Math self-concept	Negative emotions
Evaluation focus	1.00	-.20**	.38**
Math self-concept	-.20**	1.00	-.47**
Negative emotions	.38**	-.47**	1.00

Note. **Correlation is significant at the 0.01 level (2-tailed).

Table 4.4.14: Coefficients of Evaluation Focus and Math Self-concept

Model		B	SE B	β	T	Zero-order	Semi-Partial
1	(Constant)	1.76	.08		21.99***		
	Evaluation focus	.17	.02	.38	8.71***	.38	.38
2	(Constant)	3.01	.14		21.37***		
	Evaluation focus	.14	.02	.29	7.33***	.38	.29
	Math self-concept	-.38	.04	-.41	-10.32***	-.47	-.40

Note. Dependent variable: Negative emotions. ****p < .001.*

After the prerequisites for mediator testing were satisfied, the independent variable of evaluation focus alone and the independent variable of evaluation focus together with the mediator of math self-concept were used to predict the dependent variable of negative emotions. From the coefficient tables, it is observable that the direct path between evaluation focus and negative emotions became less important (zero-order correlation 0.38 decreased to 0.29 due to the mediating effect of math self-concept) but it was still significant (***Table 4.4.14)***.

Based on Sobel's (1982) solution for this complex situation, in which the direct path remains significant, though at a lower value, the method to test for a mediating relationship is to ask if the complete mediating path from independent variable to mediator to dependent variable is significant. The regression coefficients and their standard errors for the two paths in the mediating chain were shown in *Table 4.4.15*.

Table 4.4.15: Regression Coefficients and Standard Errors for Two Paths of Mediating Path from Evaluation Focus, Math Self-concept, to Negative Emotions

Path a		Path b	
Evaluation focus →Math self-concept		Math self-concept→ Negative emotions	
β_a	-0.196	β_b	-0.473
s_a	0.023	s_b	0.038
t	-4.290***	t	-11.442***

Note. ***$p < .001$

We know $t = \beta/S_\beta$. The regression coefficient for the path from Evaluation focus to Math self-concept to Negative emotions is equal to $\beta_a \times \beta_b$, where a and b refer to the relevant paths. Path c is the direct path from evaluation focus to negative emotions.

$$S_{\beta_a\beta_b} = \sqrt{\beta_a^2 s_b^2 + \beta_b^2 s_a^2 - s_a^2 s_b^2}$$

$$= \sqrt{(-0.196)^2 (0.038)^2 + (-0.473)^2 (0.023)^2 - (0.023)^2 (0.038)^2}$$

$$= 0.013$$

The mediated path coefficient was (-0.196) x (-0.473) and its standard error (0.013), therefore the t ratio was calculated as

$$t = \frac{\beta_a \beta_b}{S_{\beta_a\beta_b}} = \frac{(-0.196) \times (-0.473)}{0.013} = 7.13$$

Sobel (1982) stated that this ratio is asymptotically normally distributed, which, for large samples, would lead to rejection of the null hypothesis at a = 0.05 when the ratio exceeds ± 1.96. In this case the mediated path was clearly significant. Therefore it can be concluded that convincing evidence of a strong mediating pathway from evaluation focus through math self-concept to negative emotions existed.

- *Math self-concept as a potential mediator between parental control and negative emotions*

From the correlation table, the three paths between the independent variable of parental control and the mediator of math self-concept, between the mediator of math self-concept and the dependent variable of negative emotions, and between the independent variable of parental control and the dependent variable of negative emotions were all significantly correlated (Pearson $r = -.20$, $r = -.47$, $r = .44$, $p < 0.01$).

Table 4.4.16: Pearson Correlations among Parental Control, Math Self-concept, and Negative Emotions

	Parental control	Math self-concept	Negative emotions
Parental control	1.00	-.20**	.44**
Math self-concept	-.20**	1.00	-.47**
Negative emotions	.44**	-.47**	1.00

Note. **Correlation is significant at the 0.01 level (2-tailed).

Table 4.4.17: Coefficients of Parental Control and Math Self-concept

Model		B	SE B	β	T	Zero-order	Semi-Partial
1	(Constant)	1.75	.07		24.87***		
	Parental control	.29	.03	.44	10.29***	.44	.44
2	(Constant)	2.96	.13		22.35***		
	Parental control	.23	.03	.35	9.02***	.44	.34
	Math self-concept	-.37	.04	-.40	-10.38***	-.48	-.40

Note. Dependent variable: Negative emotions. ***$p < .001$.

After the prerequisites for mediator testing were satisfied, the independent variable of parental control alone and the independent variable of parental control together with the mediator of math self-concept were used to predict the dependent variable of negative emotions. From the coefficient tables, it is observable that the direct path between parental control and negative emotions became less important (zero-order correlation 0.44 decreased to 0.34 due to the mediating effect of math self-concept) but it was still significant (*Table 4.4.17*).

Based on Sobel's (1982) solution for this complex situation, in which the direct path remains significant, though at a lower value, the method to test for a mediating relationship is to ask if the complete mediating path from independent variable to mediator to dependent variable is significant. The regression coefficients and their standard errors for the two paths in the mediating chain were shown in the table.

Table 4.4.18: Regression Coefficients and Standard Errors for Two Paths of Mediating Path from Parental Control, Math Self-concept, to Negative Emotions

Path a		Path b	
Parental control→Math self-concept		Math self-concept→Negative emotions	
$ß_a$	-0.204	$ß_b$	-0.473
s_a	0.033	s_b	0.038
t	-4.451***	t	-11.442***

Note. ***$p < .001$

We know $t = ß/S_ß$. The regression coefficient for the path from Parental control to Math self-concept to Negative emotions is equal to $ß_a \times ß_b$, where a and b refer to the relevant paths. Path c is the direct path from parental control and punishment to negative emotions.

$$S_{ß_a ß_b} = \sqrt{ß_a^2 s_b^2 + ß_b^2 s_a^2 - s_a^2 s_b^2}$$

$$= \sqrt{(-0.204)^2 (0.038)^2 + (-0.473)^2(0.033)^2 - (0.033)^2(0.038)^2}$$

$$= 0.017$$

The mediated path coefficient was (-0.204) x (-0.473) and its standard error (0.017), therefore the t ratio was calculated as

$$t = \frac{ß_a ß_b}{S_{ß_a ß_b}} = \frac{(-0.204) \times (-0.473)}{0.017} = 5.676$$

Sobel (1982) stated that this ratio is asymptotically normally distributed, which, for large samples, would lead to rejection of the null hypothesis at $a = 0.05$ when the ratio exceeds ± 1.96. In this case the mediated path was clearly significant. Therefore it can be concluded that convincing evidence of a strong mediating pathway from parental control and punishment through math self-concept to negative emotions existed.

Summary

According to the above correlations and regression statistics, math self-concept was found to function as a mediator between positive environmental factors (teacher enthusiasm, peer attitude, and lecture engagement) and positive emotions; math self-concept was also found to function as a mediator between negative environmental factors (harsh evaluation, evaluation focus, and parental control) and negative emotions.

Therefore, the **Hypothesis 1** *(Math self-concept mediates the relationship between environmental factors and academic emotions.)* was confirmed.

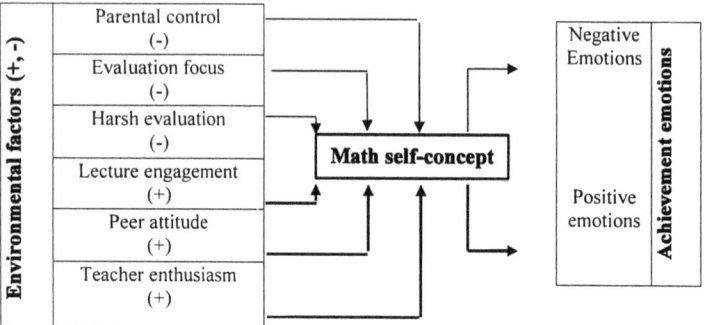

Note. (+, -) indicate the positive or negative valence of environmental factors.

Diagram 4.4.1: Math Self-concept as a Mediator on Achievement Emotions

4.4.2 Achievement goal as a mediator

Approach goal as a potential mediator between positive emotions and learning strategies

The variable of positive emotions was the mean of enjoyment and pride. The variable of approach goals was the mean of performance approach goal and mastery approach goal. The variable of learning strategies was the mean of rehearsal, elaboration, organization, critical thinking, and metacognitive self-regulation.

From the correlation table (*Table 4.4.19*), the three paths between the independent variable of positive emotions and the mediator of approach achievement goals, between the mediator of approach achievement goals and the dependent variable of learning strategies, and between the independent variable of positive emotions and the dependent variable of learning strategies were all significantly correlated (Pearson $r = .36$, $r = .35$, $r = .63$, $p < 0.01$).

After the prerequisites for mediator testing were satisfied, the independent variable of positive emotions alone and the independent variable of positive emotions together with the mediator of approach goals were used to predict the dependent variable of learning strategies. From the coefficient tables, it is observable that the direct path between positive emotions and learning strategies became less important (zero-order correlation 0.63 decreased to 0.54 due to the mediating effect of approach goals) but it was still significant.

Table 4.4.19: Pearson Correlations among Positive Emotions, Approach Goals, and Learning Strategies

	Positive emotions	Approach goals	Learning strategies
Positive emotions	1.00	.36**	.63**
Approach goals	.36**	1.00	.35**
Learning strategies	.63**	.35**	1.00

Note. **Correlation is significant at the 0.01 level (2-tailed).

Table 4.4.20: Coefficients of Positive Emotions and Approach Achievement Goals

Model		B	SE B	β	T	Zero-order	Part
1	(Constant)	1.07	.17		6.26***		
	Positive emotions	.87	.05	.63	17.01***	.63	.63
2	(Constant)	.67	.20		3.34***		
	Positive emotions	.80	.05	.58	14.79***	.63	.54
	Approach goals	.12	.03	.14	3.63***	.35	.13

Note. Dependent variable: Learning strategies.
***$p < .001$.

Table 4.4.21: Regression Coefficients and Standard Errors for Two Paths of Mediating Path from Positive Emotions, Approach Goals, to Learning Strategies

Path a		Path b	
Positive emotions⟶Approach goals		Approach goals⟶ Learning strategies	
$ß_a$	0.360	$ß_b$	0.347
s_a	0.074	s_b	0.036
t	8.199***	t	7.911***

Note. ***$p < .001$

Based on Sobel's (1982) solution for this complex situation, in which the direct path remains significant, though at a lower value, the method to test for a mediating relationship is to ask if the complete mediating path from independent variable to mediator to dependent variable is significant. The regression coefficients and their standard errors for the two paths in the mediating chain were shown in *Table 4.4.21*.

We know $t = ß/S_ß$. The regression coefficient for the path Positive emotions to Approach goals to Learning strategies is equal to $ß_a \times ß_b$, where a and b refer to the relevant paths. Path c is the direct path from positive emotions to learning strategies.

$$s_{ß_aß_b} = \sqrt{ß_a^2 s_b^2 + ß_b^2 s_a^2 - s_a^2 s_b^2}$$

$$= \sqrt{0.36^2 (0.036)^2 + 0.347^2 (0.074)^2 - (0.074)^2 (0.036)^2}$$

$$= 0.029$$

The mediated path coefficient was 0.360 x 0.347 and its standard error (0.017), therefore the t ratio was calculated as

$$t = \frac{\beta_a \beta_b}{S_{\beta_a \beta_b}} = \frac{0.360 \times 0.347}{0.029} = 4.308$$

Sobel (1982) stated that this ratio is asymptotically normally distributed, which, for large samples, would lead to rejection of the null hypothesis at a = 0.05 when the ratio exceeds ± 1.96. In this case the mediated path is clearly significant. Therefore it can be concluded that convincing evidence of a strong mediating pathway from positive emotions through approach goals to learning strategies existed.

Approach goals as a potential mediator between negative emotions and learning strategies

The variable of negative emotions was the mean of anger, anxiety, shame, hopelessness, and boredom. The variable of approach goals was the mean of performance approach goal and mastery approach goal. The variable of learning strategies was the mean of rehearsal, elaboration, organization, critical thinking, and metacognitive self-regulation.

Table 4.4.22: Pearson Correlations among Negative emotions, Approach goals, and Learning strategies

	Negative emotions	Approach goals	Learning strategies
Negative emotions	1.00	-.10*	-.48**
Approach goals	-.10*	1.00	.35**
Learning strategies	-.48**	.35**	1.00

Note. * Correlation is significant at the 0.05 level (2-tailed).
**Correlation is significant at the 0.01 level (2-tailed).

From the correlation table, the three paths between the independent variable of negative emotions and the mediator of approach achievement goals, between the mediator of approach achievement goals and the dependent variable of learning strategies, and between the independent variable of negative emotions and the dependent variable of learning strategies were all significantly correlated (Pearson r = -.10*, r = .35**, r = -.48** respectively, *p < .05 **p < .01).

After the prerequisites for mediator testing were satisfied, the independent variable of negative emotions alone and the independent variable of negative emotions together with the mediator of approach goals were used to predict the dependent variable of learning strategies. From the coefficient tables, it is observable that the direct path between negative emotions and learning strategies became less important (zero-order correlation −0.48 decreased to -0.44 due to the mediating effect of approach goals) but it was still significant.

Table 4.4.23: Coefficients of Negative Emotions and Approach Goals

Model	B	SE B	β	T	Correlations Zero-order	Part
1 (Constant)	5.58	.15		36.79***		
Negative emotions	-.70	.06	-.48	-11.46***	-.48	-.48
2 (Constant)	4.12	.24		17.56***		
Negative emotions	-.65	.06	-.45	-11.41***	-.48	-.44
Approach goals	.25	.03	.31	7.83***	.35	.31

Note. Dependent variable: Learning strategies. ***p < .001.

Table 4.4.24: Regression Coefficients and Standard Errors for Two Paths of Mediating Path

Path a		Path b	
Negative emotions → Approach goals		Approach goals → Learning strategies	
β_a	-0.097	β_b	0.347
s_a	0.084	s_b	0.036
t	-2.063*	t	7.911***

Note. ***p < .001

Based on Sobel's (1982) solution for this complex situation, in which the direct path remains significant, though at a lower value, the method to test for a mediating relationship is to ask if the complete mediating path from independent variable to mediator to dependent variable is significant. The regression coefficients and their standard errors for the two paths in the mediating chain were shown in *Table 4.4.24*.

We know $t = \beta/S_\beta$. The regression coefficient for the path from Negative emotions to Approach goals to Learning strategies is equal to $\beta_a \times \beta_b$, where a and b refer to the relevant paths. Path c is the direct path from negative emotions to learning strategies.

$$s_{\beta_a\beta_b} = \sqrt{\beta_a^2 s_b^2 + \beta_b^2 s_a^2 - s_a^2 s_b^2}$$

$$= \sqrt{(-0.097)^2 (0.036)^2 + 0.347^2 (0.084)^2 - (0.084)^2 (0.036)^2}$$

$$= 0.029$$

The mediated path coefficient was 0.360 x 0.347 and its standard error (0.017), therefore the t ratio was calculated as

$$t = \frac{\beta_a \beta_b}{s_{\beta_a \beta_b}} = \frac{(-0.097) \times 0.347}{0.029} = -1.161$$

Sobel (1982) stated that this ratio is asymptotically normally distributed, which, for large samples, would lead to rejection of the null hypothesis at a = 0.05 when the ratio exceeds ± 1.96. In this case the mediated path was clearly insignificant. Therefore it can be concluded that convincing evidence of a strong mediating pathway from negative emotions through approach goals to learning strategies did not exist.

Avoidance goals as a potential mediator between positive emotions, and learning strategies

The variable of positive emotions was the mean of enjoyment and pride. The variable of avoidance goals was the mean of performance avoidance goal and mastery avoidance goal. The variable of learning strategies was the mean of rehearsal, elaboration, organization, critical thinking, and metacognitive self-regulation.

Table 4.4.25: Pearson Correlations among Positive Emotions, Avoidance Goals, and Learning Strategies

	Positive emotions	Avoidance goals	Learning strategies
Positive emotions	1.00	-.12*	.63**
Avoidance goals	-.12*	1.00	-.02
Learning strategies	.63**	-.02	1.00

Note. * Correlation is significant at the 0.05 level (2-tailed).
** Correlation is significant at the 0.01 level (2-tailed).

From the correlation *Table 4.4.25*, only the two paths between the independent variable of positive emotions and the potential mediator of avoidance achievement goal (Pearson r = -.12*) and between the independent variable of positive emotions and the dependent variable of learning strategies were significantly correlated (r = .63**). The potential mediator of avoidance achievement goals and the dependent variable of learning strategies were not significantly correlated. Therefore the prerequisites for mediator testing were not satisfied. Since the association between avoidance achievement goals and learning strategies was not significant, there is no necessity to test the mediating role of avoidance goal between negative emotions and learning strategies.

Table 4.4.26: Pearson Correlations of Avoidance Goals, and Learning Strategies

	Avoidance goals	Rehearsal	Elaboration	Organization	Critical thinking	Metacognitive self-regulation
Avoidance goals		.15*	-.06	.06	-.11*	-.13**
Learning strategies	-.02	.77**	.87**	.77**	.78**	.83**

Note. * *Correlation is significant at the 0.05 level (2-tailed).*
** Correlation is significant at the 0.01 level (2-tailed).

In order to check which of the learning strategies were responsible for the insignificant relationship with avoidance goal, bivariat correlation was done. Avoidance goals was not significantly correlated with learning strategies in general (Pearson r = -0.02, *Table 4.4.26*). Specifically speaking, avoidance goals was positively correlated with rehearsal (Pearson r = 0.15*), negatively correlated with critical thinking and metacognitive self-regulation respectively (Pearson r = -0.11*, -0.13**). The low level significant correlations and two insignificant associations made the association between avoidance goals and learning strategies in general insignificant.

Summary

According to the above correlations and regression statistics, approach achievement goal was found to function as a mediator between positive emotions and learning strategies. Approach goal did not function as a mediator between negative emotions and learning strategies. Avoidance achievement goal did not function as a mediator between academic emotions (positive emotions and negative emotions) and learning strategies since avoidance goal was not significantly correlated with learning strategies.

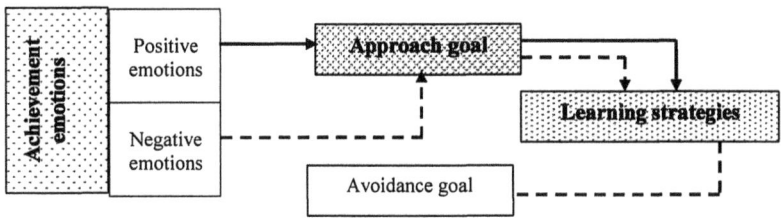

Note: The mediated path between negative emotions, approach goal, and learning strategies was eventually insignificant. Avoidance goal did not correlate with learning strategies significantly.

Diagram 4.4.2: Approach Goal as a Mediator on Learning Strategies

4.5 Regression models

4.5.1 Regressions on academic emotions
Theory base and potential Predictors
- *Potential predictors left out of the final academic emotions model*

First bivariate correlation statistic analysis was used to check the associations between potential predictors and academic emotions. According to the Pearson correlation table, social economic status (SES) was not significantly correlated with either positive emotion or negative emotion, therefore SES was not put into the regression model.

Table 4.5.1: Pearson Correlations of Emotions and Potential Predictors

	Self-concept	Self-efficacy	Intrinsic value	Extrinsic value	Lecture engagement	Harsh evaluation	Evaluation focus	Parental control	Peer attitude	Teacher enthusiasm	SES	GPA	Attitude
Positive emotion	.50**	.50**	.64**	.26**	.65**	-.28**	-.39**	-.20**	.48**	.42**	-.07	.47**	.64**
Negative emotion	-.47**	-.48**	-.46**	.01	-.56**	.30**	.38**	.44**	-.27**	-.36**	.09	-.52**	-.58**

Note. $**p < .01$ *(2 tailed)*

Based on the statistic results obtained from MANOVA and Tukey test on achievement emotions by achievement and gender (*Table 4.2.9 & Table 4.2.10*), differences on achievement emotions were not found by gender. Therefore, gender was not put into the regression model.

- *Potential predictors put in the final academic emotions model*

Based on the previous statistical analysis on academic emotions, it has been found out that math self-concept functioned as a mediator between the environmental factors and academic emotions. Therefore, math self-concept was regarded as an important potential predictor in the final academic emotions model.

Based on the social cognitive control-value theory (Pekrun, 2000), The following potential predictors were put in the final models of academic emotions (positive emotions model and negative emotions model), which were appraisal factors (math self-concept, self-efficacy in math learning, and values), environmental factors (lecture engagement, harsh evaluation, evaluation focus, parental control, peer attitude, teacher enthusiasm), Math GPA, and attitude toward math. Achievement goals were also loaded in the models. Among the numerous potential predictors, which of them could really predict the criterion variable of academic emotions, which of them had the relative strongest prediction power were meaningful questions to be explored.

Regression on the model of positive academic emotions

According to the model summary, the best model explained 60 percent variances for the dependent variable of positive emotions. The best model was found to be significant (F (7, 350) = 76.04 at .001 level).

The best model included the following significant predictors: attitude toward math class (ß = 0.21, t = 4.20***), lecture engagement (ß = 0.18, t = 3.57***), math self-concept (ß = 0.18, t = 4.37***), mastery approach (ß = .14, t = 3.75***), peer attitude (ß =0.15, t = 3.76***), intrinsic value in math learning (ß = 0.14, t = 2.91**), and math GPA (ß = 0.11, t = 2.76**). It can be concluded that positive affect in math learning is often associated with high control-value beliefs and positive environmental factors.

Negative environmental factors, e.g., parental control and punishment, harsh evaluation, evaluation focus, and several other factors, namely, self-efficacy, extrinsic value in math learning, teacher enthusiasm, were not picked up as significant predictors by the regression model.

Table 4.5.2: Multiple Regression Analysis for Variables Predicting Positive Academic Emotions

Variable	B	SE B	ß	t
Attitude towards math	.15	.04	.21	4.20***
Lecture engagement	.10	.03	.18	3.57***
Math self-concept	.17	.04	.18	4.37***
Mastery approach	.09	.02	.14	3.75***
Peer attitude	.11	.03	.15	3.76***
Intrinsic value	.13	.04	.14	2.91**
Math GPA	.003	.001	.11	2.76**

Note. $F (7, 350) = 76.04***$, $Adj. R^2 = .60$.
$**p < .01$. $***p < .001$.

Regression on the model of negative academic emotions

By using the forward multiple regression method, the best model explained 61 percent variances for the dependent variable of negative emotions. The best model was found to be significant (F (9, 348) = 61.74 at .001 level).

The best model included the following predictors: attitude toward math class (ß = -0.27, t = -5.72***), performance avoidance (ß = .22, t = 5.92***), math GPA (ß = -.12, t = -2.95**), math self-concept (ß = -.18, t = -4.55***), evaluation focus (ß = .09, t = 2.17*), lecture engagement (ß = -.18, t = -3.65***), parental control and punishment (ß = .10, t = 2.59*), peer attitude (ß = .10, t = 2.46*), and harsh evaluation (ß = .08, t = 2.05*). Five of the nine predictors were from the environmental factor category. It can be concluded that negative affect in math learning is often associated with low control-value beliefs and to a large extent influenced by environmental factors.

Table 4.5.3: Multiple Regression Analysis for Variables Predicting Negative Academic Emotions

Variable	B	SE B	β	t
Attitude toward math	-.19	.03	-.27	-5.72***
Performance avoidance	0.9	.02	.22	5.92***
Math GPA	-.003	.001	-.12	-2.95**
Math self-concept	-.17	.04	-.18	-4.55***
Evaluation focus	.04	.02	.09	2.17*
Lecture engagement	-.09	.03	-.18	-3.65***
Parental control	.07	.03	.10	2.59**
Peer attitude	.07	.03	.10	2.46*
Harsh evaluation	.06	.03	.08	2.05*

Note. $F(9, 348) = 61.74^{***}$, $Adj. R^2 = .61$.
$^*p < .05.$ $^{**}p < .01.$ $^{***}p < .001.$

Summary

By comparing the model for predicting positive academic emotions and model for predicting negative academic emotions, the common predictors but with opposite valences for both models were attitude toward math class, math self-concept, math GPA, and lecture engagement. In addition, mastery approach was one of the strongest predictors for positive emotions and performance avoidance was one of the strongest predictors for negative emotions. The above findings confirmed the associations between appraisals of ability and emotions, and environmental factors and emotions.

4.5.2 Regressions on learning strategies

Theory base and potential predictors

According to social cognitive control-value theory (Pekrun, 2000), learning process, specifically learning strategy use has direct association with emotions. Learning strategy has further indirect associations with appraisals and environmental factors. Appraisals mainly refer to control beliefs (self-concept, self-efficacy) and value appraisals (intrinsic value and extrinsic value). Environmental factors include distal (culture and society), and proximal (parents, teachers, and peers) factors. In order to test learning strategy and its associations with the direct and indirect factors, a multiple regression model on learning strategy was constructed based on the following potential predictors: emotions (enjoyment, pride, anger, anxiety, shame, hopelessness, and boredom), appraisals (math self-concept, math self-efficacy, intrinsic value, extrinsic value), proximal environmental factors (parental control, peer attitude, teacher enthusiasm, classroom climate and its subfactors of lecture engagement, harsh evaluation, evaluation focus).

Since achievement goals might also contribute to learning strategy use, the four achievement goals (performance approach, mastery avoidance, mastery approach, performance avoidance) were also put into the final regression model. Based on statistic results obtained from MANOVA and Tukey test on learning strategy use by achievement and gender (*Table 4.2.9 & Table 4.2.10*), both achievement and gender predicted learning strategy use in math learning, therefore math GPA and gender were put into the regression model as well. Attitude toward math was put in the final model of learning strategies, too.

Regression on the model of learning strategies

Model summary revealed the best fit model explained about 48 percent variances for the dependent variable of learning strategy use. The best fit model based on the seven predictors was significant (F (7, 350) = 47.70 at the .001 level).

Table 4.5.4: Multiple Regression Analysis for Variables
Predicting Learning Strategy Use

Variable	B	SE B	β	t
Enjoyment	.33	.09	.25	3.58***
Mastery approach	.18	.04	.21	4.66***
Harsh evaluation	.20	.04	.19	4.61***
Boredom	-.22	.06	-.21	-3.54***
Pride	.23	.07	.19	3.35**
Math self-concept	.16	.06	.12	2.68**
Extrinsic value	-.12	.05	-.10	-2.42*

Note. F (7, 350) = 47.70***, Adj. R² = .48.
*p < .05. **p < .01. ***p < .001.

The regression model for learning strategy use (*Table 4.5.4*) was predicted by enjoyment (β = 0.25, t = 3.58***), mastery approach (β = 0.21, t = 4.66***), harsh evaluation (β = 0.19, t = 4.61***), boredom (β = -.21, t = -3.54***), pride (β = .19, t = 3.35**), math self-concept (β = 0.12, t = 2.68**), and extrinsic value (β = -.10, t = -2.42*). Among the seven predictors, five were positive predictors with boredom and extrinsic value being the only two negative predictors, which means experiencing frequently boredom and having strong extrinsic value predicted using less learning strategies in learning mathematics.

Summary

The results confirmed learning strategy and its strong association with emotions. Positive affect positively predicted learning strategy use; negative affect negatively predicted learning strategy use. Math self-concept was a strong predictor for achievement emotions (*Table 4.5.2 & Table 4.5.3*). In the current model math self-concept was again found to predict learning strategy use in learning mathematics significantly.

Compared with parental control, peer attitude, and teacher enthusiasm, harsh evaluation (one of the subfactors of classroom climate) was a more immediate predictor for learning strategies. Although parental control, peer attitude, and teacher enthusiasm were proximal environment like classroom climate, which, however, was the real learning and working environment students were in and therefore had stronger impact on students' learning strategy use. Mastery approach and extrinsic value (negative predictor) were found to predict learning strategy use, too. However, neither math GPA nor gender predicted learning strategies within this model.

4.5.3 Regressions on math achievement

Theory base and potential predictors

According to social cognitive, control-value theory of achievement emotions (Pekrun, 2000), academic achievement is influenced by motivation and learning strategy use directly, and indirectly influenced by achievement emotions. Appraisals, namely self-concept, self-efficacy, and values, mediate environmental factors' impact on achievement emotions. Environmental factors' impact might reach academic achievement.

In order to test the theory, the following variables were loaded into the regression model. Environmental factors loaded into the model were parental control, peer attitude, teacher enthusiasm, and perceived classroom environment and its subfactors: lecture engagement, harsh evaluation, and evaluation focus. Appraisal predictors loaded into the model were math self-concept, self-efficacy, intrinsic value, and extrinsic value in math learning. Achievement emotion predictors loaded into the model were enjoyment, pride, anger, anxiety, shame, hopelessness, and boredom. Learning strategy predictors loaded into the model were rehearsal, elaboration, organization, critical thinking, and metacognitive self-regulation. Achievement goal predictors loaded into the model were performance approach, performance avoidance, mastery approach, and mastery avoidance. Attitude toward math and gender were put into the regression model, too.

Regression on the model of math achievement

The model summary indicated that there were seven significant predictors for math GPA and the predictors explained the variances for the criterion variable of math GPA at a rather high level (R square = .45, adjusted R square = .44).

Table 4.5.5: Multiple Regression Analysis for Variables Predicting Math Achievement

Variable	B	SE B	β	t
Boredom	-8.21	1.69	-.27	-4.85***
Parental control	-6.68	1.22	-.25	-5.48***
Math self-concept	5.34	1.76	.14	3.04**
Performance approach	2.19	.71	.13	3.11**
Metacognitive self-regulation	5.29	1.59	.20	3.34***
Rehearsal	-2.84	1.10	-.12	-2.59**
Harsh evaluation	-2.65	1.33	-.09	-2.00*

Note. $F(7, 350) = 40.71***$, Adj. $R^2 = .44$.
* $p < .05$. ** $p < .01$. *** $p < .001$.

MANOVA statistics further confirmed the significance of the math GPA regression model (F (7, 350) = 40.71 at the .001 level), which included the following significant predictors: boredom (ß = -.27, t = -4.85***), parental control (ß = -.25, t = -5.48***), math self-concept (ß = .14, t = 3.04**), performance approach (ß = .13, t = 3.11**), metacognitive self-regulation (ß = .20, t = 3.34***), rehearsal (ß = -.12, t = -2.59**), and harsh evaluation (ß = -.09, t = -2.00*).

Boredom, parental control, rehearsal, and harsh evaluation were negative predictors for math GPA, which means strong associations existing between high level of boredom and low math achievement, strong parental control and low math achievement, frequent use of basic cognitive strategy of rehearsal and low math achievement, and harsh evaluation and low math achievement.

Summary

The statistical results obtained from the regression models on emotions, learning strategies, and math GPA were in line with the social cognitive, control-value theory of achievement emotions (Pekrun, 2000). Math achievement was predicted by performance approach goal (positive predictor in the category of motivation), rehearsal (negative predictor in the category of learning strategies), and metacognitive self-regulation (positive predictor in the category of learning strategies). Math achievement was also predicted by boredom (negative predictor in the category of academic emotions). Harsh evaluation and parental control (negative predictors in the category of environmental factors) were predictors of math achievement. Math self-concept (positive predictor in the category of appraisal) predicted math achievement, too.

In line with the previous research (Pekrun et al., 2002a) it was found that (test) anxiety was not necessarily the most detrimental negative academic emotion. Specifically, text anxiety proved to relate less closely to achievement than hopelessness and boredom. From the theories concerning affect in learning and especially anxiety literature, it is known that anxiety has multi-functions. Up to a certain high level for each individual anxiety could be facilitative in terms of accelerating cognitive and metacognitive processes. However, over that specific high point for each individual, anxiety could be detrimental since it burdens cognitive and metacognitive processes instead of accelerating the thinking and monitoring processes, detracts attention from tasks, and even inhibits cognitive and metacognitive processes. The tolerant anxiety levels are variant among individuals.

Achievement values, attitude toward math, and gender were not found to be significant predictors for math achievement by the regression model.

4.5.4 Integrated picture of math learning processes and achievement model

On primary variables

- *On self*

Math self-concept and self-efficacy were loaded in all the four regression models together with other potential predictors. Math self-concept predicted positive academic emotions, learning strategies and math achievement positively, and negative academic emotions negatively (*Table 4.5.6*). Besides being a strong predictor, math self-concept was found to mediate the path between environmental factors and academic emotions (*see Diagram 4.4.1 Math Self-concept as a Mediator on Achievement Emotions*).

- *On achievement values*

Achievement values were loaded in all the regression models together with other potential predictors. Intrinsic value in math learning predicted positive emotions positively. Extrinsic value in math learning predicted learning strategies negatively (*Table 4.5.6*). However, neither intrinsic value nor extrinsic value predicted math achievement or negative emotions.

Therefore, a prediction that "*Intrinsic value is more facilitative than extrinsic value in math learning process.*" made previously (p. 152) concerning the contribution of achievement values in math learning process was confirmed.

- *On environmental factors*

Environmental factors were loaded in all the regression models with other potential predictors. Environmental factors included perceived classroom climate (lecture engagement, harsh evaluation, and evaluation focus), parental control, peer attitude, and teacher enthusiasm.

From the summary (*Table 4.5.6*), positive environmental factors positively predicted positive emotions and negatively predicted negative emotions; negative environmental factors positively predicted negative emotions, and eventually negatively predicted math GPA. Therefore, it could be concluded that negative environmental factors are detrimental to math learning processes and outcome.

- *On achievement goals*

Achievement goals were loaded in the emotions, learning strategies, and math GPA

Table 4.5.6: Summary of Predictors for Academic Emotions, Learning Strategies and Math GPA

	Academic Emotions		Learning strategies	Math GPA
	Positive emotions Model 1	Negative emotions Model 2	Model 3	Model 4
Self [1-4]				
Math self-concept	X	X -	X	X
Self-efficacy				
Values [1-4]				
Intrinsic value	X			
Extrinsic value		X -		
Environmental factors [1-4]				
Lecture engagement	X	X -		
Harsh evaluation		X	X	X -
Evaluation focus		X		
Parental control		X		X -
Peer attitude	X	X		
Teacher enthusiasm				
Academic emotions [3-4]				
Enjoyment			X	
Pride			X	
Anger				
Anxiety				
Shame				
Hopelessness				
Boredom			X -	X -
Attitude toward math [1-4]	X	X -		
Math GPA [1-3]	X	X -		
Gender [3-4]				
Achievement goals [1-4]				
Performance approach				X
Mastery avoidance				
Mastery approach	X		X	
Performance avoidance		X		
Learning strategies [4]				
Rehearsal				X -
Elaboration				
Organization				
Critical thinking				
Metacognitive self-regulation				X

Note: X means significant predictor. "–" indicates the negative valence of the predictor.
"1 - 4" means predictors for Model 1 to Model 4.
"3 - 4" means predictors for Model 3 to Model 4.
"1 - 3" means predictors for Model 1 to Model 3.
"4" means predictors only for Model 4.
The order of the predictors is not their prediction magnitude.

models together with other potential predictors. Mastery approach goal predicted both positive emotions and learning strategies use. Performance avoidance goal predicted negative emotions. Mastery goals disappeared in the math achievement model. Only performance approach goal predicted math achievement (*Table 4.5.6*). Harackiewicz and Elliot also found that performance-approach goal was beneficial in terms of achievement (1998).

- *On academic emotions*

Academic emotions were loaded in learning strategies, and math achievement models together with other potential predictors. From the summary (*Table 4.5.6*), positive affect predicted more learning strategies use, and negative affect predicted less learning strategies use. However, enjoyment and anxiety should be analysized in broadened theory contexts. Negative affect also predicted low math GPA. Therefore, it could be concluded that negative affect is not only detrimental to learning processes but learning outcome as well.

- *On learning strategies*

Learning strategies were loaded in the math achievement model together with other potential predictors. Basic cognitive strategy, namely rehearsal was found to predict math achievement negatively, and metacognitive self-regulation was found to predict math achievement positively (*Table 4.5.6*).

- *On attitude toward math, math GPA, and gender*

Attitude toward math was loaded in all the regression models. Attitude was found to predict affective processes, namely positively predicted positive emotions and negatively predicted negative emotions. It did not predict math GPA or learning strategy use.

Math GPA was loaded in the emotions and learning strategies models. Math GPA was found to predict affective processes.

Gender was loaded in the learning strategies and math achievement models together with other potential predictors. Gender did not predict learning strategy use or math GPA (*Table 4.5.6*).

Joint regression analyses

Regression analyses were conducted below from a more general perspective, namely associations were analysized based on categorized variables rather on specific variables. Technically, specific environmental factors were combined into positive environmental factor (average of lecture engagement, peer attitude, and teacher enthusiasm) or negative environmental factor (average of parental control, harsh evaluation, and evaluation focus); specific achievement emotions were sorted into positive emotion category (average of pride and enjoyment) or negative emotion category (average of anger, anxiety, hopelessness, shame, and boredom); specific achievement goals were grouped as approach goal (average of performance approach goal and mastery approach goal) or avoidance goal (average of performance avoidance goal and mastery avoidance goal).

Joint regression analyses aimed at presenting the interrelations in a structured way and the integrated picture visually. *Figure 4.5.1* is a pictorial summary of the primary results.

- *Environmental factors, math self-concept, and achievement values as predictors of academic emotions*

When the environmental factors, self-concept, and achievement values were tested together as joint predictors of positive emotions ($R^2 = .56$, $F(5, 449) = 114.71$, $p < .001$), each of the joint variables were shown to be a significant predictor variable: intrinsic value ($t = 6.70$, $p < .001$, $\beta = .29$), positive environment ($t = 8.22$, $p < .001$, $\beta = .34$), math self-concept ($t = 6.37$, $p < .001$, $\beta = .23$), extrinsic value ($t = 2.48$, $p < .05$, $\beta = .08$), negative environment ($t = -1.97$, $p < .05$, $\beta = -.07$).

When the environmental factors, self-concept, and achievement values were tested together as joint predictors of negative emotions ($R^2 = .45$, $F(5, 448) = 73.97$, $p < .001$), each of the joint variables were shown to be a significant predictor variable: negative environment ($t = 6.84$, $p < .001$, $\beta = .28$), math self-concept ($t = -6.45$, $p < .001$, $\beta = -.26$), positive environment ($t = -3.90$, $p < .001$, $\beta = -.18$), intrinsic value ($t = -4.23$, $p < .001$, $\beta = -.20$), extrinsic value ($t = 3.84$, $p < .001$, $\beta = .15$). See *Figure 4.5.1* for a pictorial summary of the primary results.

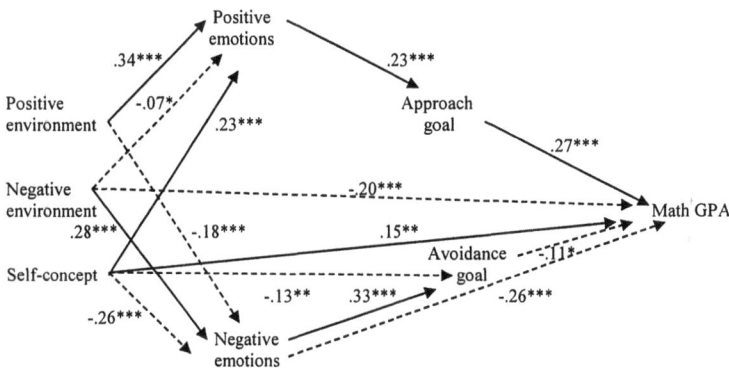

Note. Path values are standardized regression coefficients. Only the theoretically central variables are included in the diagram for presentation clarity. *$p < .05$. **$p < .01$. ***$p < .001$.

Figure 4.5.1: A Summary of the Results from Joint Regression Analyses

- *Environmental factors, math self-concept, achievement values, and emotions as predictors of achievement goals*

When the environmental factors, self-concept, achievement values, and academic emotions were tested together as joint predictors of approach goals ($R^2 = .35$, $F(2, 449) = 122.89$, $p < .001$), only extrinsic value and positive emotions were shown to be significant predictor variables: extrinsic value ($t = 12.47$, $p < .001$, $\beta = .49$), positive emotions ($t = 5.85$, $p < .001$, $\beta = .23$).

When the environmental factors, self-concept, achievement values, and academic emotions were tested together as joint predictors of avoidance goals ($R^2 = .31$, $F(3, 448) =$

67.70, p < .001), only negative emotions, extrinsic value and math self-concept were shown to be significant predictor variables: negative emotions (t = 7.46, p < .001, β = .33), extrinsic value (t = 9.74, p < .001, β = .38), math self-concept (t = -2.92, p < .01, β = -.13). See *Figure 4.5.1* for a pictorial summary of the primary results.

- *Environmental factors, math self-concept, achievement values, emotions, and goals as predictors of math GPA*

When the environmental factors, self-concept, achievement values, emotions, and goals were tested together as joint predictors of math GPA (R^2 = .36, F(5, 352) = 41.76, p < .001), the following significant predictor variables were identified: negative emotions (t = -4.41, p < .001, β = -.26), approach goals (t = 5.08, p < .001, β = .27), negative environment (t = -3.98, p < .001, β = -.20), math self-concept (t = 2.92, p < .01, β = .15), avoidance goal (t = -2.02, p < .05, β = -.11). See *Figure 4.5.1* for a pictorial summary of the primary results.

Stepwise regression at each major step in the analysis showed that the addition of each successive set of predictor variables increased the total explained variance of math achievement (see *Figure 4.5.2*).

Positive environment predicted positive emotions; however, it did not predict approach goal orientation or math achievement. Negative environment predicted negative emotions and math achievement. Neither approach goal nor avoidance goal was influenced directly by environmental factors. Math self-concept predicted both positive and negative emotions, avoidance goal, and math achievement. Positive emotions impacted math achievement indirectly through approach goal; whereas negative emotions influenced avoidance goal and math achievement directly. Both approach goal and avoidance goal worked on math achievement directly. Math achievement was directly dangered by negative environmental factor, negative affect, and avoidance goal orientation.

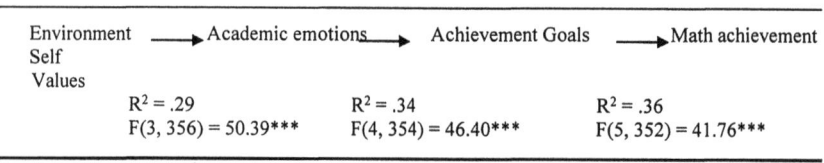

Note. ***p < .001. R^2s are adjusted R^2s.

Figure 4.5.2: Stepwise Regression for Math Achievement

Dynamic affective-cognitive learning processes and achievement model

Based on all the results, learning processes (appraisal, affective experiences, learning strategies use, and achievement goals) and learning outcome (math achievement) were a dynamic cycle, which was influenced by environmental factors. The term "dynamic" in the current context has three levels of meaning. Level 1 – math learning outcome could be improved through multiple means. Focusing on learning strategies alone is not enough. Level 2 – variables in this model are modifiable on the condition that other relevant variables are modified for the purpose of reaching desired outcome, e.g., increased positive affect in math learning leading to more efficient learning strategy use and self-regulation. Level 3 – learning outcome and achievement is by no means the end product, which further influences learners' emotions in learning, goal orientation, learning strategy use in math learning.

The big dynamic cycle was then constructed by sub-dynamic cycles (*Diagram 4.5.1*): centering on affective experiences, goal orientations, learning strategies and self-regulation, and learning outcome respectively. In the sub-dynamic cycle centering on affective experiences, self-concept mediated environmental impact on affective experiences. Affective experiences were mainly predicted by environmental factors, self-concept, and achievement values. In the sub-dynamic cycle centering on goal orientation, goal orientation was mainly predicted by self-concept, values, and affective experiences. In the sub-dynamic cycle centering on learning strategies and self-regulation, learning strategies and self-regulation was predicted by environmental factors, self-concept, values, affective experiences and goal orientation. When goal orientation was approach, approach goal mediated positive emotions' impact on learning strategies but did not mediate negative emotions' impact on learning strategies; when goal orientation was avoidance, avoidance goal did not function as a mediator between affective experiences and learning strategies.

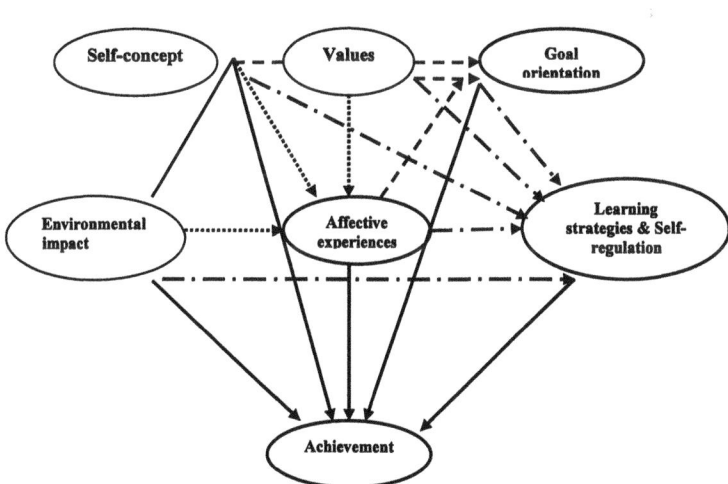

Diagram 4.5.1: Affective-Cognitive Learning Processes and Achievement Model

In the sub-dynamic cycle centering on learning outcome, learning outcome (*Table 4.5.5*) was contributed by achievement goal (performance approach goal: ß = 0.13, t = 3.11, at .01 level), learning strategies use (rehearsal: ß = -.12, t = -2.59, at .01 level and metacognitive self-regulation: ß = .20, t = 3.34, at .001 level), affective experiences (boredom: ß = -.27, t = -4.85 at .001 level), appraisal (math self-concept: ß = .14, t = 3.04 at .01 level), and proximal environmental factors (harsh evaluation: ß = -.09, t = -2.00 at .05 level and parental control: ß = -.25, t = -5.48, at .001 level). In the current study, achievement values, attitude toward math, and gender did not predict math GPA.

Chapter V CONCLUSIONS AND DISCUSSIONS

5.1 Factor structure of the measures among the Chinese sample

In general, the Chinese sample had similar factor structures of learning strategy use, achievement emotions, achievement goals and perception of classroom climate like those of western samples in previous studies (e.g., Pintrich et al., 1991; Elliot & McGregor, 2001; Church, Elliot, and Gable, 2001).

5.1.1 Learning strategies

The Chinese sample's use of learning strategies were loaded on six factors through exploratory factor analysis using principal axis factoring with promax rotation. Critical thinking and some of the metacognitive items were loaded as one factor. Cognitive and metacognitive items coexisted in factor 2, 3, and 4. Factor 5 and 6 were composed of all together 3 metacognitve items. The above factor structure of the learning strategies gave support that critical thinking and metacognitive were close to each other although some metacognitive strategies were loaded on cognitive strategy scales. The mixture of metacognitive and cognitive especially cognitive complex strategies further reflected that the functions of some complex cognitive strategies were somehow approaching that of metacognitive strategies. The factor validity for the MSLQ within the Chinese sample was reasonable and sound.

Pintrich et al. (1991) tested the factor validity of the Motivated Strategies for Learning Questionnaire through confirmatory analyses (LISREL VI Joreskog & Sorbom, 1986). They provided several fit statistics: the chi-squared to degrees of freedom ratio (χ^2/df); the goodness of fit index (GFI; generated by the LISREL VI program); the root mean residual (RMR); and Hoelter's critical number (CN). A χ^2/df ratio of less than 5 is considered to be indicative of a good fit between the observed and reproduced correlation matrices (Hayduk, 1987). A GFI of .9

or greater; an RMR of .05 or less; and a CN of 200 and above are heuristic values that indicate that the models "fits" the input data well (Pintrich et al., 1991). Pintrich et al.'s fit statistics are the following: χ^2/df = 2.26; GFI = .78; RMR = .08; CN = 180 and they claimed that the fit statistics were reasonable values and the factor validity for the MSLQ learning strategy scale was sound (Pintrich et al., 1991) although three of the four fit statistics were not sound enough. In sum, the MSLQ was a valid tool to measure Chinese students' use of learning strategies.

5.1.2 Achievement emotions

The Chinese sample's achievement emotions were loaded on seven factors using maximum likelihood with promax rotation. The chi-square to degrees of freedom ratio (χ^2/df) was 1.85. A χ^2/df ratio of less than 5 is considered to be indicative of a good fit between the observed and reproduced correlation matrices (Hayduk, 1987). Shame, hopelessness, and anxiety items were found to coexist in Factor 3, Factor 4, Factor 5, which further confirmed the reality that the three emotions were inseparable or closely connected with one another in math learning context in China. The three emotions were inwardly directed, reflecting limited resources to cope with outward demand. Ability was found to be the most salient predictor for anxiety, shame, and hopelessness (*Table 4.3.11/12/13*). Anger and boredom were loaded on one factor and inherently different from shame, hopelessness, and anxiety. Anger and boredom were emotions mainly directing outward. Teaching quality and task difficulty were found to be responsible for anger and boredom (*Table 4.3.10/14*). Enjoyment and Pride, the two positive emotions, were loaded on one factor and were separated from the rest negative emotions.

The factor structure of the academic emotions confirmed the existence of positive and negative emotions in learning mathematics, and further differentiated the inwardly directing negative emotions from the outwardly directing negative emotions. Therefore, the AEQM was a valid tool to measure Chinese students' emotions in learning mathematics.

5.1.3 Achievement goals

The Chinese sample's achievement goals were loaded on three factors using principal-axis factoring with promax rotation. The three factors were mastery goal, performance avoidance goal, and performance approach goal. The mastery approach goal and mastery avoidance goal of the Chinese sample were loaded as one factor of mastery goal, which means the distinctiveness of the two mastery goals disappeared within the Chinese sample. The two performance goals were exactly the same as those of western samples in terms of item numbers and item quantities. Therefore, it gave support that the two performance goals existed within the Chinese sample. The similar factor structure within the Chinese sample reveals that achievement goals exist across cultures and the Achievement Goal Questionnaire (Elliot & McGregor, 2001) was a valid tool to measure Chinese students' goal orientations in learning mathematics.

5.1.4 Perceived classroom climate

The Chinese sample's perceived classroom climate was loaded on three factors using principal-axis factoring with promax rotation. The three factors were lecture engagement, evaluation focus and harsh evaluation. The three factors represent separable perceived classroom environment constructs although two of the harsh evaluation items were loaded on evaluation focus subscale. The lecture engagement factor was the same as that of the western sample (Church, Elliot, and Gable, 2001) in terms of item numbers and item quantities. Thus, the factors of perceived classroom environment found within the Chinese sample were congruent with the factor construct of the original measure and Perceived Classroom Climate Questionnaire (Church, Elliot, and Gable, 2001) was usable with a sample from a collectivist culture.

5.2 Math, a gender or achievement issue

A child begins learning mathematics in his or her primary school days and this learning process lasts for his or her whole life span in various contexts, e.g., school and work, etc.. The higher grades the child is in, the more demanding mathematics becomes. Besides cognitive challenges, unfortunately, this learning process always results in unavoidable evaluations and normative comparisons with others because of its essential role in school and work. Consequently, students are not only cognitively but emotionally involved in math learning process as well. Their values and self-concept are challenged, too. Sadly, only students' math performance is highlighted and paid extraordinary attention on whereas their cognitive effort even pain, emotional disturbs even sufferings, and vulnerable self-concept are ignored or not given enough attention to.

The achievement or in another word normative issue becomes more complicated with a learner's gender. Males are believed to have higher ability in this domain than females. Teachers and parents expect males to achieve more than females in mathematics. In the masculine domain of math, being a female seems to be a born obstacle to be good at math. Is it really the case that females are inferior than males in learning mathematics and females suffer more from learning mathematics than males?

A scientific view of sorting out the projected question is to investigate the motivational aspects and affective experiences of male and female learners besides their math performance and achievement. In addition, specific aspects involved in learning mathematics should be compared among comparable groups in order to reach reasonable conclusions, that is to say, statistical results from pure gender comparisons do not help much catch the differences among a specific gender group, namely differences between high achievement males and low achievement males, and differences between high achievement females and low achievement females. For example, MANCOVA statistics when controlling prior math achievement, males

(Mean = 3.01, SD = .72) were found to have more positive math self-concept (F (2, 347) = 37.23, p < .001) than females (Mean = 2.75, SD = .62) (*Table 4.2.7*). However, when subjects' achievement background was considered besides their gender, differences (Mean difference$_{Female\ groups}$ = .42, p < .05; Mean difference$_{Low\ achievement\ groups}$ = -.30, p < .05; *Table 4.2.5*) were only identified between high achievement females (Mean = 3.03, N = 92) and low achievement females (Mean = 2.60, N = 131), and low achievement females (Mean = 2.60, N = 131) and low achievement males (Mean = 2.90, N = 91) respectively (*Table 4.2.3/5*). High achievement females and males did not differ on math self-concept and males did not differ on math self-concept no matter they were in high achievement group or in low achievement group. Obviously the four group post hoc comparison has shown us more meaningful information than two gender comparison.

Only in this way, can a learner's natural born characteristics, motivation, learning habits and performance be reasonably comprehended and understood. An exclusive association between gender and math achievement contains too little meaningful information since gender does not necessarily lead to certain level of math achievement. According to the current study, males (Mean = 112.41, N = 185) were found to have higher math achievement (F(1, 361) = 8.46, p < .01) than females (Mean = 104.47, N = 178). Does this mean that gender predicts math achievement? Based on the current study, gender was not even one of the predictors for math achievement since the other predictors (boredom, parental control, math self-concept, performance approach, metacognitive self-regulation, rehearsal, harsh evaluation) were more powerful to predict math achievement than gender ($\beta_{Boredom}$ = -.27, t = -4.85, p < .001; $\beta_{Parental\ control}$ = -.25, t = -5.48, p < .001; $\beta_{Math\ self-concept}$ = .14, t = 3.04, p < .01; $\beta_{Performance\ approach}$ = .13, t = 3.11, p < .01; $\beta_{Metacognitive\ SR}$ = .20, t = 3.34, p < .001; $\beta_{Rehearsal}$ = -.12, t = -2.59, p < .01; $\beta_{Harsh\ evaluation}$ = -.09, t = -2.00, p < .05) (*Table 4.5.5*). The above regression evidence further gave support that gender effect should not be overestimated and exaggerated but be considered in a pool of factors working on math achievement.

The results from univariate, multivariate, and post hoc statistics shows that the relevant aspects of math learning should not be attributed to gender or achievement level simply but should be analysized under the interaction of the two. In addition, different aspects are influenced by different mechanisms, which are even beyond the issues of gender and achievement.

5.2.1 A trial to understand learners' cognitive functioning – learning strategies use

Analysis of learners' use of learning strategies is a channel to understand one important aspect of learners' learning habits. Is use of learning strategies solely explained by gender or achievement or both? According to the statistical results of the current study, use of learning strategies was explained by both gender and achievement. Specifically, elaboration, metacognitive self-regulation, and critical thinking differences were found by achievement, in another word, high achievers applied more elaboration (Mean difference $_{Female}$ = .49, p < .05; Mean difference $_{Male}$ = .27, p < .05), metacognitive self-regulation (Mean difference $_{Female}$ = .82, p < .05; Mean difference $_{Male}$ = .55, p < .05) and critical thinking (Mean difference $_{Female}$ = .91, p < .05; Mean difference $_{Male}$ = .67, p < .05) strategies than low achievers; high achievement females applied more rehearsal (Mean difference = .43, p < .05) and organization strategies (Mean difference = .50, p < .05) than high achievement males; low achievement females applied less critical thinking strategy (Mean difference = -.60, p < .05) than low achievement males (*Table 4.2.9*).

Based on the above post hoc statistics, learning strategy use differed by both gender and achievement level. However, according to the multiple regression statistics on use of learning strategies, neither gender nor achievement predicted use of learning strategies since affect, mastery approach, math self-concept, etc., were stronger predictors for learning strategies (*Table 4.5.4*). Therefore, the intuitive thinking that girls apply less learning strategies than boys in math learning process is not accurate.

5.2.2 Affective experiences in learning mathematics

Analysis of affective experiences in math learning helped us answer the question: does females suffer more and enjoy less in math learning process than males? The reasoning for post hoc group difference statistics have been clarified under 5.2. A pool of factors work on learners' affective experiences, therefore it is too general to just consider gender impact on affective experiences. The Tukey test gave convincing support that it was not accurate to claim females suffered more and enjoyed less in math learning than males; in addition, low achievers suffered more and enjoyed less than high achievers of the same gender (low achievement females suffered more and enjoyed less than high achievement females, low achievement males suffered more and enjoyed less than high achievement males); furthermore, gender differences of affective experiences across achievement levels (high achievement females vs. high achievement males, low achievement females vs. low achievement males) were not found.

Among the negative affective experiences, experiencing shame was not different among the four achievement by gender groups (Mean $_{High\ achievement\ male}$ = 2.51, SD = .60; Mean $_{High\ achievement\ female}$ = 2.51, SD = .58; Mean $_{Low\ achievement\ male}$ = 2.53, SD = .58; Mean $_{Low\ achievement\ female}$ = 2.56, SD = .61) (*Table 4.2.9*). In another word, experiencing shame is such a frequent and common experience in math learning process in China no matter the learner is a high achiever or low achiever, male or female. The mechanism for this finding could be collectivist culture, which is constructed on affiliative needs and one has to constantly strive not to make one's family lose face. Experiencing shame frequently could be an indicator of one's conscious reflection of one's mission: not to disappoint one's family and teacher and therefore to try harder.

In sum, four (gender by achievement) group comparisons was a more heuristic tool to look into affective experiences in learning mathematics than two (gender) group comparisons since less variance explaining affective experiences was lost. According to the post hoc statistics, variance of affective experiences in math learning was explained by learners' achievement level rather than by gender and therefore the intuitive thinking that girls suffer more and enjoy less in math learning process is not accurate.

5.2.3 Who is more motivated to learn?

Analysis of learners' achievement goal orientation helps us answer the question: who is more motivated to learn? Could it be that males are more motivated than females in math learning since math has been regarded as a masculine domain? Results from post hoc statistics (Tukey test) showed that achievement motivation was significantly different by achievement level rather than by gender. Specifically, high achievement females (Mean $_{\text{Performance approach}}$ = 5.17, SD = 1.45; Mean $_{\text{Mastery approach}}$ = 6.48, SD = .69) had higher level of performance approach (Mean difference = .56, p < .05) and mastery approach (Mean difference = .69, p < .05) than low achievement females (Mean $_{\text{Performance approach}}$ = 4.59, SD = 1.67; Mean $_{\text{Mastery approach}}$ = 5.79, SD = 1.28), high achievement males (Mean $_{\text{Mastery approach}}$ = 6.19, SD = .86) had higher level of mastery approach (Mean difference = .71, p < .05) than low achievement males (Mean $_{\text{Mastery approach}}$ = 5.48, SD = 1.27); high achievers both females (Mean = 3.91, SD = 1.56) and males (Mean = 3.56, SD = 1.51) had lower performance avoidance (Mean difference $_{\text{Female}}$ = -.90, p < .05; Mean difference $_{\text{Male}}$ = -.70, p < .05) than low achievers, namely low achievement females (Mean = 4.81, SD = 1.38) and males (Mean = 4.26, SD = 1.44) respectively (*Table 4.2.9* & *Table 4.2.10*).

Difference of mastery avoidance goal was not found either by achievement or gender. In another word, the concern of not being able to grasp all math knowledge motivated high and low achievement female and male learners in a similar way.

Therefore, variance of achievement motivation was explained by learners' achievement level rather than by their gender and the intuitive thinking that boys are more motivated than girls in math learning is not plausible.

5.2.4 Confidence in learning mathematics

In the masculine domain of math, are boys more confident and do they have more control than girls, namely do boys have more positive math self-concept and higher level of self-efficacy than girls? In Eccles' (1983) study to investigate students' intention to take additional math courses, females, in comparison with males, had lower confidence in their ability and perceived math as more difficult and less valuable. Frenzel et al. (2007b) also found that females had lower confidence in their ability than males when their prior math achievement was controlled. Eccles (1983) even found that such sex differences appeared to be related to parents' beliefs in the difficulty of math for their child. However, Eccles only compared male and female math learners in general and did not consider learners' achievement background.

The results from post hoc (Tukey test) statistics in the current study showed that confidence difference was found with low achievers across gender (Mean difference $_{\text{Self-concept}}$ = .30, p < .05; Mean difference $_{\text{Self-efficacy}}$ = .47, p < .05); high achievement males and females' self-concept and self-efficacy were not significantly different. In addition, it was found that high achievement females had more positive self-concept (Mean difference = .42, p < .05) and higher level of self-efficacy (Mean difference = .49, p < .05) than low achievement females (*Table 4.2.5*). Interestingly, low achievement males and high achievement males' math self-concept and self-efficacy were not significantly different. Parents' expectations for high achievement males and low achievement males were also not significantly different. It is an important issue to be further explored in the future whether parents' expectations backup children's confidence in academic work and even a step further what factors shape parents' high expectations for boys.

Therefore, variance of confidence was explained by both achievement and gender and the intuitive thinking that boys are more confident than girls in learning math is not accurate.

5.2.5 Parental expectancy for boys and girls

Self achievement expectations of boys and girls according to the current study were congruent with their achievement level; in another word high achievers had higher self achievement expectation than low achievers. It was not the case with parental expectancy, which had a particular gender preference for males. Parental expectancy for low achievement boys was higher than that for low achievement girls (Mean difference = .43, p < .005) although the two groups had similar low math performance. The gender preference disappeared in high achievement male and female groups, which means girls' math achievement must be really high and convincing enough to buffer parental expectancy for them, whereas parental expectancy for boys was always high no matter they were high or low achievers (Mean difference = .30, p < .089) (*Table 4.2.5*). The mechanism that girls' high math achievement buffered parental expectancy functioned again between high achievement female and low achievement female groups, namely parental expectancy for high achievement girls were higher than low achievement girls (Mean difference = .40, p < .05. *Table 4.2.5*).

Therefore, parental expectancy favors boys and is unconditionally high no matter boys are high or low achievers; parental expectancy for girls is contingent upon girls' math achievement and it is only high when girls' math achievement is competitively high. In a collectivist culture, filial piety and fulfilling one's parents' wishes is the major reason for achievement motivation. High parental expectancy could function as a positive force to push up one's motivation to achieve. Thus, boys have an advantage over girls in terms of "unconditional parental expectancy".

5.2.6 Summary

By so far, the issue involved in research question three, namely whether variances of various aspects in math learning are decided by gender or achievement, has been discussed.

III. **(Research question)** Do high achievers' cognition-emotion processes differ from low achievers' processes? Do male students' cognition-emotion processes differ from female students' processes?

A heuristic view toward the debate is that math learning is a complex process involving cognitive, affective and motivational aspects to be evaluated first individually and then jointly to reach an integrated picture. The intuitive thinking that boys are more cognitively advanced in terms of applying learning strategies, enjoy more and suffer less, are more motivated to learn math, and have more confidence in learning math were challenged.

Boys were found to have an advantage over girls in terms of unconditional parental expectancy, which functions as a positive external force to push boys to achieve, whereas girls had to be really high achievers to gain high parental expectancy. However, boys' advantages over girls were few.

Differences in cognitive functioning in terms of using learning strategies were mainly found between high achievers and low achievers; across gender difference was only found between low male and female achievers on critical thinking. Affective experiences and achievement motivation were only found by achievement. Difference of math learning confidence was found between males and females of low achievement groups. High achievement girls and boys did not differ on applying elaboration, critical thinking, and metacognitive self-regulation strategies, affective experiences, achievement motivation, and confidence in learning mathematics.

Based on the above findings, the intuitive and stereotype thinking that boys are cognitively more advanced, enjoy more and suffer less, are more motivated to learn, and are more confident in learning math than girls could not be proved to function at both high and low achievement levels. Therefore, it is hazardous to turn the intuitive thinking into inferences on affective-cognitive processes in learning mathematics across genders. A lot of meaningful information would be lost if only gender and math learning outcome were highlighted.

5.3 Interrelations of affective-cognitive processes

Research questions one and two focused on interrelations of affective-cognitive processes, which were analysized through looking into the correlations of the relevant affective and cognitive variables. The correlations between environmental factors and affective-cognitive processes were also considered besides the interrelations of affective-cognitive processes.

I   (**Research question**) What are the dynamics of adolescents' cognition-emotion processes, namely the interrelations between self-concept, achievement values, emotions and motivation, in the achievement-related context of math learning?

II   (**Research question**) How do social and individual antecedents associate with adolescents' academic emotions and achievement motivation?

5.3.1 Associations of affective-cognitive variables

The general pattern was that positive affect and motivational components associated with adaptive cognitive functioning and high math achievement; negative affect and motivational components associated with maladaptive cognitive functioning and low math achievement.

Specifically, learning strategies positively correlated with positive affect and negatively correlated with negative affect. Control beliefs (self-concept and self-efficacy) positively correlated with positive affect negatively correlated with negative affect. At the same time, control beliefs positively correlated with learning strategies use. Intrinsic value's associations with affect and learning strategies were the same as control beliefs; however, extrinsic value was positively correlated with a few negative emotions. Negative affect (boredom: ß = -.27, t = -4.85 at .001 level), math self-concept (ß = .14, t = 3.04 at .01 level), learning strategies (metacognitive self-regulation: ß = .20, t = 3.34, at .001 level and rehearsal: ß = -.12, t = -2.59, at .01 level) were found to predict mathematics achievement (*Table 4.5.5*).

Students' focus on a mastery goal was found to be related with a wide range of important motivational, cognitive, and affective outcomes. Specifically, mastery goal orientation was positively related with self-efficacy, use of effective learning strategies, positive school-related affect. These findings were consistent with previous findings (e.g., Ames & Archer, 1988; L. Anderman, 1999; Kaplan & Midgley, 1999; Midgley, Anderman, & Hicks, 1995). However, performance approach goal (ß = 0.13, t = 3.11, at .01 level) was found to predict math achievement (*Table 4.5.5*). This finding is also consistent with previous findings (Church et al., 2001).

5.3.2 Associations between environmental factors and affective-cognitive variables

Positive environmental factors positively correlated with positive affect, learning strategy use, control beliefs, intrinsic value and approach goals. At the same time, negative environmental factors positively correlated with negative affect and avoidance goals, negatively correlated with learning strategy use, control beliefs, and intrinsic value. Among the environmental factors, lecture engagement was positively correlated with intrinsic value (r = .61, p < .001), whereas harsh evaluation (r = -.16, p < .001) and evaluation focus (r = -.34, p < .001) were negatively correlated with intrinsic value (*Table 4.3.1*). The findings concerning perceived classroom environment were consistent with previous findings (Church et al., 2001).

Negative environmental factors (parental control: ß = -.25, t = -5.48, at .001 level and harsh evaluation: ß = -.09, t = -2.00 at .05 level) were found to predict math achievement (*Table 4.5.5*). However, based on the current study positive environmental factors were not predictors for math achievement. According to the joint regression analyses (*Table 4.5.1*), the impact of positive environmental factors on math achievement were mediated by positive emotions and approach goal successively. From these findings, it can be concluded that negative environmental factors were particularly detrimental to students' emotional well-being, motivation in learning, cognitive functioning, math self-concept and eventually math achievement and no aspects of affective-cognitive processes could mediate negative environmental factors' detrimental impact on achievement.

Therefore, parents should give their children more autonomy over their study instead of exercising too much control on them; teachers are suggested to lower their evaluation harshness and make it more acceptable by students since extreme harsh evaluation would hurt students' self-concept and motivation tremendously; evaluation focus could be oriented toward mastering tasks when a certain level of excellence has been achieved. In sum, providing students with supportive and friendly learning environment is one of the channels to accommodate students' emotional needs and facilitate adaptive learning pattern, which are prerequisites for excellent performance.

5.4 From environmental factors to emotions and from emotions to learning strategies

The mechanisms of environmental factors on emotions and emotions on learning strategies were the focal interests of research question four, which were analysized through mediator testing, namely whether math self-concept could function as a mediator between environmental factors and emotions and whether achievement goal could function as a mediator between emotions and learning strategies.

IV. (**Research question**) How does learning environment influence students' affective experiences and how do students' affective experiences influence learning strategy use?

5.4.1 Math self-concept's mediating effect on emotions

Math self-concept was found to mediate positive environmental factors' (lecture engagement, peer attitude, teacher enthusiasm) impact on positive emotions. It was also found to mediate negative environmental factors' (parental control, normative evaluation focus, harsh evaluation) impact on negative emotions (*Diagram 4.4.1*). The mediating effect suggested that environmental factors' impact on achievement emotions were not direct but mediated by math self-concept, in another word this impact was reduced because of math self-concept's contribution to achievement emotions. The mediating effect is unavoidable since math self-concept is an inherent part of a child's identity.

What does math self-concept mediating positive environmental factors' impact on positive emotions mean exactly? It means that the zero-order correlations between positive emotions and teacher enthusiasm (Pearson r = .42 at .01 level *Table 4.4.1*), between positive emotions and peer attitude (Pearson r = .48 at .01 level *Table 4.4.4*), and between positive emotions and lecture engagement (Pearson r =.65 at .01 level *Table 4.4.7*), were higher than they actually were. It also means that when the effect of math self-concept was considered the partial correlations between positive environmental factors and positive emotions would become weaker.

The practical implications based on the mediating mechanism could be that students' positive emotions would not rise with improved positive environmental factors constantly since students' math self-concept is rather stable and it prevents positive emotions from inflating with improved learning environment. In a very friendly learning environment (high teacher enthusiasm, positive peer attitude toward math, high lecture engagement), a realistic math self-concept is especially important to stabilize a student's positive affect within a reasonable range.

What does math self-concept mediating negative environmental factors' impact on negative emotions mean exactly? It means that the zero-order correlations between negative emotions and parental control (Pearson r = .44 at .01 level *Table 4.4.16*), between negative emotions and normative evaluation focus (Pearson r = .38 at .01 level *Table 4.4.13*), between negative emotions and harsh evaluation (Pearson r = .30 at .01 level *Table 4.4.10*), were higher than they actually were. It also means that when the effect of math self-concept was considered the partial correlations between negative environmental factors and negative emotions would become weaker.

The practical implications based on the mediating mechanism could be that students' negative emotions would not rise with worsened negative environmental factors constantly since students' math self-concept is rather stable and it prevents negative emotions from inflating with worsened learning environment. In another word, students' math self-concept buffers and serves as a protective barrier against unfriendly learning environment's (high parental control, normative evaluation focus, harsh evaluation, etc.) negative influence on students' emotional well-being.

5.4.2 Achievement goals' mediating effect on learning strategies

Approach goal was found to mediate positive emotions' impact on learning strategy use. Approach goal did not mediate negative emotions' impact on learning strategy use since the mediated path was eventually insignificant. Avoidance goal did not mediate emotions' impact on learning strategy use since avoidance goal did not correlate with learning strategies significantly (*Diagram 4.4.2*). The mediating effect suggested that positive emotions' impact on learning strategy use was not direct but mediated by approach goal, in another word this impact was reduced because of approach goal's contribution to learning strategies.

What does approach goal mediating positive emotions' impact on learning strategy use mean exactly? It means that the zero-order correlations between positive emotions and learning strategy use (Pearson r = .63 at .01 level *Table 4.4.19*) was higher than it actually was. It also means that when the effect of approach goal was considered the partial correlation between positive emotions and learning strategies would become weaker.

The practical implications based on the mediating mechanism could be that students' learning strategy use would not be efficient with frequent experiencing positive affect in math learning without the coexisting of approach goal in the relation. Although an approach goal's mediating effect on learning strategies is important, experiencing negative emotions frequently could not be mediated by approach goal. In another word, use of learning strategies would not be efficient and effective if students experience too much negative affect in learning mathematics even though they are trained to use approach goals. Another implication could be increasing students' positive affective experiences in math learning and this issue has been clarified in 5.4.1 Math self-concept's mediating effect on emotions.

5.5 Dynamic affective-cognitive learning processes and outcome model

5.5.1 Math achievement: a multidimensional construct

Dynamic affective-cognitive learning processes and outcome model is presented in *Diagram 4.5.1*. According to the multiple regression analysis (*Table 4.5.5*), math achievement was found to be a multidimensional construct and its variances were explained by various affective-cognitive and environmental factors, namely affect (boredom as a negative predictor), learning environment (parental control and harsh evaluation as negative predictors), math self-concept, achievement goal (performance approach), and learning strategies (rehearsal as a negative predictor and metacognitive self-regulation as a positive predictor). Gender was not found to predict math achievement since the above predictors were more powerful than gender.

Based on the above findings, there are many hints to educators who are dedicated to help students adopt adaptive way of learning, and many hints to students who want to improve their math achievement. It was the initial purpose of the researcher to look for and provide practical hints to educators, parents, and students.

The predictors for math achievement are from two resources: either internal, e.g., math self-concept, performance approach goal, and boredom, or external, e.g. parental control, and harsh evaluation. However, internal variables are not absolutely decided internally but are under the influence of external variables, which are reciprocally influenced by internal variables. For example, a child's strengthened achievement motivation would lead to changes in the child's learning environment, e.g., decreased parental control. The improved facilitative learning environment would then cause the child to experience less negative emotions, to identify himself or herself with a more positive math self-concept, and even apply more effective learning strategies in math learning process. Consequently, the adaptive way of learning would aid the child to achieve better in the field of mathematics. The merit of dynamic affective-cognitive processes and outcome model is best reflected in the intervention model for improving learning outcome (*Diagram 5.6.1*).

5.5.2 Implications for teachers, parents, and students in brief

In such a dynamic learning cycle, what an educator can intervene is to make the learning environment more supportive, e.g., giving less harsh evaluation, not to hurt students self-esteem, to be sensitive to students' emotions in learning, etc. For example, boredom could be a dangerous signal that a child is not interested at all in learning and achievement.

Home is also an important learning environment besides school and some aspects of home can be changed more supportive for children, e.g., parents giving more autonomy to children over their own learning and trying to exercise less control over their children. More autonomy is a prerequisite for a child to become a real self-regulated learner and then to achieve more.

A child should learn to appreciate the virtue of effort, to learn math through approach goals and to apply more effective and efficient learning strategies in learning. The ultimate goal for a child is to become a self-regulated learner to learn efficiently and effectively and consequently to achieve a certain excellence. Self-regulating refers to regulating one's own positive and negative emotions (adopting a realistic self-concept, controlling and transforming one's negative emotions to achievement motivation, etc.), regulating one's learning strategy use, and regulating one's learning environment (avoiding destructive learning environment and looking for facilitative learning environment, etc.).

5.6 Pedagogical implications: multiple channels to support development and enhance achievement

5.6.1 Formula of development

As Piaget (1952) has put it, organism and environment form an entity. Maturation, according to Piaget, is the organism's fundamental tendency to organize experience so that it can be assimilated; learning is the means of introducing new experiences into that organization. To fully understand the mechanism of development is a prerequisite to support development.

According to Piaget (1952), maturation (nature) interacts with learning (nurture) to form development. It is important to note that they are not additive (+) but interactive (X). The formula is

>Maturation X Learning = Development
>In the absence of experience of a specified sort, the equation becomes
>Maturation X Zero learning = Zero achievement
>Another formula applies in the absence of maturation:
>Zero maturation X Learning opportunity = Zero achievement.

The final task of adolescence is to complete the integration of the developmental changes (biological, social, psychological, cognitive) into a well-adjusted young adult (Jones, 1992). Cognitive changes in the adolescent include the acquisition of abstract thought processes, logical reasoning, hypothesizing, and metacognitive thinking skills, which enable adolescents to deal with relatively complex mathematics and math related subjects (Jones, 1992b). Becoming aware of emotionality changes is one of the psychological changes happening during adolescence (Haviland, Gebelt, & Stapley, 1997). Getting in touch with one's feeling and emotions is a sign of maturity (Haviland, Gebelt, & Stapley, 1997, p.246). Being aware of one's feelings and emotions is helpful for emotion regulation. Due to adolescents' cognitive, psychological readiness, providing them with necessary learning opportunities, e.g., introducing social skills to enhance emotional competence, is beneficial to their development.

5.6.2 Enhancing emotional competence through introducing social skills

The notion of competence has been defined as the capacity or ability to engage in transactions with a variable and challenging social-physical environment, resulting in growth and mastery for the individual (White, 1963). Emotional intelligence represents the core aptitude or ability to reason with emotions. Emotional achievement represents the learning a person has attained about emotion or emotion-related information, and emotional competence exists when one has reached a required level of achievement. All things being equal, a person's emotional intelligence determines her emotional achievement. But things are rarely equal, and the family in which one grew up, the lessons about emotions one was taught, the life events one has undergone – all influence how much one has achieved in learning about emotions (Mayer & Salovey, 1997, p.15). The social skills introduced below can enhance children/adolescents' emotional competence and social competence.

When emotional competency is viewed from a more comprehensive perspective of social competence, the following social skills can be taught to kids: impulse control, anger management, empathy, recognizing similarities and differences among people, complimenting, self-monitoring, communication, evaluating tasks, positive self-talk, problem solving, decision making, goal setting, etc.(Defalco, 1997, p.33).

5.6.3 Maintaining and enhancing self-concept

For the adolescent, self-concept development involves constructing a self-portrait which integrates the different characteristics of the self. This process includes realistically evaluating his/her intellectual competence, physical competence, physical attractiveness, social competence, leadership abilities, moral beliefs, and sense of humor (Jones, 1992).

For the adolescent, the self-concept now functions as a standard for evaluating and predicting performance socially, objectively, personally; and functions to limit performance for the purpose of maintaining and enhancing itself (Manaster, 1989). "Compared to younger children, early adolescents are highly self-conscious and have uncertain, shaky images themselves with regard to certain qualities they value" (Skolnick, 1986, 463). Adolescents are able to maintain a reasonably positive self-image by identifying positive attributes as core constructs in their self-portrait and considering negative attributes or behaviors as foreign to their true self (Harter, 1986).

From the current study, math self-concept was found to be in the dynamic cycle of learning processes and outcome and furthermore it predicted positive emotions positively, negative emotions negatively, learning strategy use positively, and math achievement positively. On the one hand, self-concept of one's ability is highly important in adolescents' affective well-being and school achievement, on the other hand, it is pretty vulnerable under external influences. Therefore, maintaining and enhancing adolescents' self-concept is an essentially challenging task for adolescents, teachers, and of course parents. The concrete methods to maintain and enhance self-concept could be: parents and teachers giving more autonomy to students over their own study and exercising less control, showing respect to students' decisions on their study, giving less criticism, evaluating students' academic performance reasonably, etc..

5.6.4 Enhancing achievement motivation from a goal perspective

Linnenbrink (2002) claimed that personal and contextual goals may have unique influences on various learning outcomes. Personal performance-approach goals were unrelated to adaptive motivation, cognitive strategy use, and help-seeking; detrimental for social loafing and achievement; and beneficial for emotional well-being. While mastery context condition was beneficial for adaptive help-seeking and emotional well-being, it was also associated with lower achievement. Furthermore, the performance context condition was detrimental to help-seeking behaviors and emotional well-being and was associated with higher levels of achievement. In addition, Linnenbrink (2004) did not believe that personal goals functioned as a mediator between context goals and learning outcomes although some others did (e.g., Maehr, Pintrich, and Zimmerman, 1993; Anderman and Maehr, 1994). In sum, contextual goals and personal achievement goals influence learning outcomes. Thus, enhancing achievement motivation and learning outcomes from a goal perspective is possible. Before clarifying concrete means to enhance learning outcomes it is necessary to have a look at motivation in the middle grades.

Motivation in the middle grades

Motivation is important at all stages. However, issues of motivation have a degree of uniqueness and certainly a special sense of urgency about them during the middle grades since motivation's direct outcome, namely achievement may determine career trajectories. A number of studies have indicated that, during the middle grades, students often exhibit a disturbing downturn in motivation (Anderman & Maehr, 1994), self-concept of ability, and positive attitudes toward school decrease, particularly during grades six and seven (Harter, 1981; Marsh, 1989). Declines in motivation during adolescence are associated with contextual/environmental factors and that motivation is not merely a function of pubertal changes (Eccles & Midgley, 1989; Simmons & Blyth, 1987). A number of motivational researchers (e.g., Eccles et al., 1993a; Maehr & Midgley, 1991; Weinstein & Butterworth, 1993) suggest that differences in the instructional practices and educational policies between elementary and middle schools often are inappropriate for maintaining the motivation of students after the transition.

- *Developmental changes in terms of cognitions*

As students approach adolescence, they tend to view ability more as a stable, internal trait and as less related to effort than they did earlier (e.g., Dweck & Leggett, 1988; Nicholls, 1986, 1989). The inference in the situation of a test failure is likely to be that one is incompetent. A defensive maneuver employed by many students to avoid this judgment is not to study (Anderman & Maehr, 1994). Students will use such techniques to avoid failure and maintain a sense of self-worth (Covington, 1992). Some researchers suggest that self-concept and self-esteem change during early adolescence as a function of changes in home and school environments (e.g., Eccles et al., 1993b). Further, as children move through childhood into early adolescence, their self-concepts include more psychological descriptors in addition to behavioral and physical descriptors and are based more strongly on information received from social comparisons with other children (Wigfield & Karpathian, 1991).

- *Changes in terms of context and achievement goal adoption*

Eccles and her colleagues (Eccles et al., 1993b) argue that there is a developmental mismatch between the psychological needs of early adolescents and the types of environments that most schools provide. The typical middle school environment is characterized by little autonomy, excessive rules and discipline, homogeneous grouping by ability and stricter evaluation practices compared with those in the elementary school years (Eccles et al., 1993a; Eccles & Midgley, 1989; MacIver & Epstein, 1993). However, early adolescence is best nurtured by a strong sense of autonomy, independence, self-determination, and social interaction (Carnegie, 1989; Eccles & Midgley, 1989). A mismatch between the context which typical middle schools provide for early adolescents and the psychological needs of youth exists.

Middle school climate is normally ability goal oriented, which means middle school students are trained to demonstrate their ability or outperform others, whereas elementary school environments stress task-focused goals (Ames, 1992b; Ames & Archer, 1988). A large literature suggests that these goals are orthogonal and not simply opposite ends of a continuum (Maehr & Pintrich, 1991). These goals can have qualitatively different effects on many types of behaviors (Maehr, Pintrich, & Zimmerman, 1993). Students who have task-focused goals and focus on task mastery are very likely to be influenced by a school climate focusing on performance and ability comparison since the school as a whole can influence the goals that students adopt (Maehr, 1991; Maehr, Medgley, & Urdan, 1992).

Building a task-focused learning environment: intervention

Achievement goal theory assumes that achievement goal structures can change or influence students' personal goal orientation (Ames, 1992b). The idea that changing the context can influence students' own goals makes achievement goal theory particularly useful in school reforms efforts, as specific aspects of the classroom or school environment can be altered to accommodate and influence students' own goal adoption and learning (Linnenbrink, 2004). Psychological environment is a precursor for the personal goal beliefs that individuals hold. Self-efficacy mediates motivation differentially depending on what goal dominates. The goals that students adopt have been shown to be related to cognitive strategies, achievement, behaviors, and affect (Maehr et al., 1993; Urdan & Roeser, 1993).

Ames (1990, 1992a) promoted a mastery goal structure, which emphasized six related categories: the academic Task, Authority, Recognition, Grouping, Evaluation, and Time (e.g. TARGET, see also Epstein, 1989). Patrick (2004) elaborated the six categories. *Tasks* should be meaningful, challenging, and interesting. The teacher should share *authority* and responsibility for rules and decisions with the students. *Recognition* should involve progress or effort, and there should be few opportunities for social comparison among students. *Grouping* should be heterogeneous. *Evaluation* should not be made public and grades should be interpreted in terms of improvement. And there should be flexible use of *time* in the classroom. TARGET has featured prominently in recommendations about creating more adaptive, mastery-focused learning environments (e.g., Maehr & E. Anderman, 1993; Maehr & Midgley, 1991; Midgley, 1993; Midgley & Edelin, 1998; Midgley et al., 2002; Midgley & Urdan, 1992).

Ryan and Patrick (2001) further found that the dimensions of teacher support, mutual respect, and promoting task-related interaction in the social context of the classroom may be an inextricable part of what it means to have a classroom emphasis on developing competence, or a mastery goal structure. Aspects of the classroom social environment are closely related to, or integral components of, a classroom mastery goal structure (Patrick, 2004).

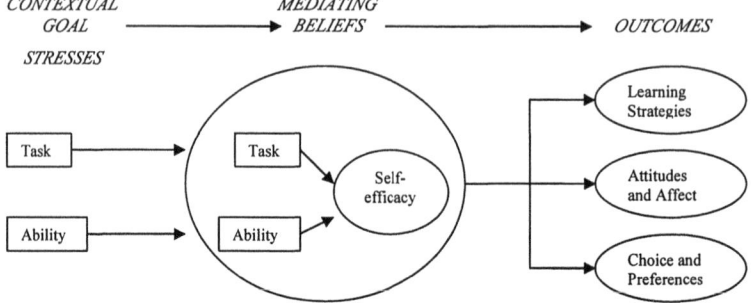

Note: Adapted from Maehr, Pintrich, and Zimmerman (1993) by Anderman and Maehr (1994).

Figure 5.6.1: Schematic Representation of Goal Theory Model

The guidelines for reform are explicit, direct, and practical but also somewhat revolutionary in nature. Carried to their implied ends, they suggest that motivation change in the middle grades will eventuate in the transformation of school culture (Sarason, 1990).

5.6.5 Intervention model for improving learning outcome

The intervention model for improving learning outcome was based on the empirical findings from the current study and previous studies, and relevant social-cognitive achievement motivation and emotion theories. Special attention could be given by teachers to the encouragement of perceived competence in academic endeavors, to the promotion of autonomous behavior, and to the validation of the student as an active partner in a shared learning experience designed to increase both intrinsic motivation and positive affect, making learning more enjoyable (Csikszentmihalyi, 1997; Renninger, 2000). The teacher can create conditions leading to an upward spiral of positive growth for students by minimizing extrinsic pressure and maximizing curiosity, interest, and flow in the classroom (Fredrickson & Losada, 2005; Kashdan & Fincham, 2004). Humor, respect, and social support nurture students' academic experience in the classroom (Bergin, 1999), which creates an opportunity to turn a domain-specific interest into an individual interest in learning (Alexander, Murphy, Woods, Duhon, & Parker, 1997). According to Pekrun et al. (2002a), some general guidelines for the designing of optimal instructional environment are: (a) improve the quality of instruction; (b) give students autonomy, but only to the extent that they are able to self-regulate their learning; (c) convey high values of academic mastery, but adjust social expectations for achievement so that they match students' capabilities; (d) increase opportunities for success by using individual and cooperative reference norms for giving feedback and by inducing a culture of learning from errors; and (e) create flexible interaction structures that foster affiliation, cooperation, and support without denying the role of competition among peers (Astleitner, 2000; Covington, 1992; Pekrun, 1998; Perry, Schonwetter, Magnusson, & Struthers, 1994).

The intervention for improving learning outcome is organized and focused on two levels with maintaining and enhancing self-concept as the focus of the level one and increasing positive affect and decreasing negative affect as the focus of the level two. The level one intervention should be carried out before the level two since the level one is more or less the foundation for the level two intervention. The two intervention levels have all together six intervention areas (*Diagram 5.6.1*).

5.6.5.1 Intervention level I: Maintaining and enhancing self-concept as the focus in companion with motivational scaffolding

The mechanism of the intervention level one is accommodating learners' needs to improve person environment fit so that learners' self-concept is preserved, maintained, and even enhanced. Research in a wide range of academic contexts attests to the importance of enhancing efficacy beliefs to achieve substantive outcomes, such as increased levels of academic performance in mathematics (Schunk, 1983, 1984; Schunk & Cox, 1986), and general academic performance (Bandura, 1997; Lent, Brown, & Gore, 1997; Multon, Brown, & Lent, 1991; Zimmerman, Bandura, & Martinez-Pons, 1992).

However, it is cautious to note that providing supportive learning environment is not the only means to maintain and enhance learners' academic self-concept, which is shaped by learners' academic performance, parental expectation, etc., as well. Supportive learning environment has the following elements: focusing tasks and mastering, improving the cognitive quality of instruction, giving learners more autonomy over their learning pace, content, exercising less normative evaluation especially public evaluation, and less harsh evaluation. Unnecessary criticisms are suggested to be avoided for their counterproductive effects on motivation and affect in students (Deci & Ryan, 2000). Concerning autonomy, extrinsically motivated students tended to report more negative affect than intrinsically motivated students under autonomy-supportive conditions, even when students' self-concept was controlled; for directive parental support, the reverse trend was discoved. Positive emotions are evoked when a match between motivation and situation is perceived; negative emotions are evoked when a mismatch between motivation and situation is perceived. Intrinsically motivated students need autonomy-supportive support to realize optimal learning processes, whereas extrinsically motivated students need directive parental support to realize optimal learning outcome instead of optimal learning processes (Knollmann & Wild, 2007).

Therefore, the intervention model for improving learners' psychological well-being and learning outcome needs to work on learners' internal psychological world, e.g., motivation, self-concept, affect, etc., and learners' external learning environment, and in addition needs to pay attention to the congruence/incongruence between the interventions on internal psychological world of motivation and external learning environment. Furthermore, the person-environment fit model of motivation and instruction of Knollmann and Wild (2007) can be used to foster a change from extrinsic to intrinsic forms of motivation since parents adapt their support style to their child's extrinsic motivation, positive emotions are evoked; in turn, these emotions foster the development of intrinsic motivation. Therefore, motivational scaffolding is only possible when parents are able to detect their children's current motivation as well as the proximal zone of motivational development (Brophy, 1999) and adjust their support accordingly. The supportive learning environment is for the benefit of building positive academic self-concept. The supportive environment and activated intrinsic value are like the two wings of self-concept. Without the two wings the intervention level one cannot be fully realized and the intervention level two would lose its foundation.

5.6.5.2 Intervention level II: Emotion regulation as the focus in companion with cognitive scaffolding

The mechanism of the intervention level two is emotion regulation, namely increasing learners' positive affect and decreasing negative affect in learning processes based on maintained and enhanced self-concept plus activated intrinsic value. Typically, but not always, emotion regulation strives to increase positive emotions, and decrease negative emotions (coping with negative emotions; Zeidner & Endler, 1996). Basic components of emotion regulation are recognition and understanding one's own emotions, managing these emotions by inducing, modulating, or preventing them, and using emotions for action and goal attainment (Matthews, Zeidner, & Roberts, 2002). A similar definition of emotion regulation was given by Gross (1998) who suggested that emotion regulation is "the process by which individuals influence which emotions they have, when they have them, and how they experience and express these emotions" (p.275). Emotion regulation involves various processes such as monitoring, evaluating, and modifying one's emotional process (Schutz & Davis, 2000; Sutton, 2004).

Emotions are central to psychological health and well-being, implying that they should be regarded as educational outcomes in themselves, independent of their functional relevance (Pekrun, 2006). From the perspective of functional relevance, positive affect is so important for learners to adopt adaptive learning method, taking approach goals and exercising active self-regulation and using efficient learning strategies, etc.. Approach goals and active self-regulation are the two wings of emotion regulation, which should more or less lead to improved psychological well-being and learning outcome. Teaching and encouraging use of learning strategies and self-regulation is essentially cognitive scaffolding. Teachers need to be aware of students' current cognitive capacity as well as the proximal zone of cognitive development in order to optimize cognitive scaffolding.

Although positive emotions and absence of intense negative emotions are core elements of well-being, positive emotions can sometimes be detrimental for important outcomes like achievement, and negative emotions can be beneficial. Thus, from the perspective of outcome attainment and future well-being, the pattern is more complex than simplistic hedonism would suggest (Pekrun, 2006). Therefore, increasing positive affect and decreasing negative affect could be regarded as looking for the reasonable proportion of positive and negative affect along the dimensions of emotion regulation rather than the ultimate goal of approaching the ends of the two dimensions in learning contexts. Emotion regulation should eventually facilitate learners' self-regulation of emotions based on control-value theory.

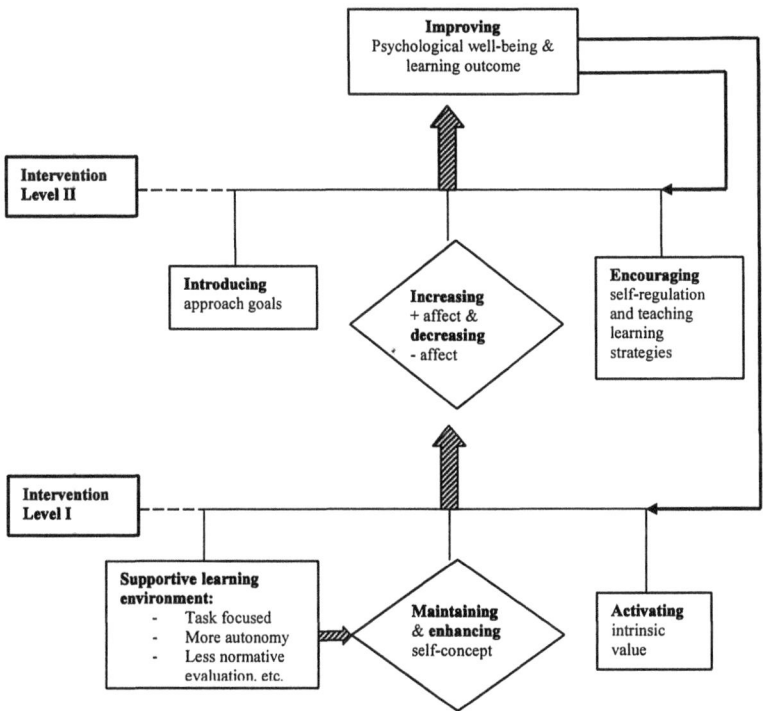

Diagram 5.6.1: Intervention Model for Improving Psychological Well-Being and Learning Outcome

The improved psychological well-being and learning outcome is the product of the two interventions but it does not stop there. It reciprocally enhances learners' self-concept, increases positive affect, and decreases negative affect. The assumption of the model is that emotions, learning, and achievement are linked by reciprocal causation, with students' self-concept, emotions, and motivation influencing achievement, and feedback of achievement in turn affecting their self-concept, emotions and motivation. The interventions – outcome 1 – the reciprocal enhancing effect on interventions – outcome 2 – ... , namely a dynamic cycle is the purpose of the intervention.

5.7 Limitations of the study and future research

The current study is mainly cross-sectional in nature, with the data being collected through a comprehensive questionnaire survey. Correlation-regression statistics besides analyses of variances, exploratory factor analysis, etc., were applied to catch the differences of affective and cognitive variables by achievement and gender groups, and the associations between and among the variables. The purpose of understanding adolescents' affective-cognitive processes in learning mathematics has been so far fairly achieved but the generalizability of the findings is limited due to its cross-sectional nature, namely no change or cause-effect relationships could be convincingly inferred and presented.

A concrete example could be: according to the intercorrelation analysis between emotions and learning strategies (*Table 4.3.1*), positive emotions positively correlated with learning strategies and negative emotions negatively correlated with learning strategies except for the complication of experiencing shame. However, the conclusion that positive affect leads to use more learning strategies and negative affect leads to use less learning strategies could not be made. In order to make causal effect conclusions, field experimental study manipulating students' affect in learning and observing consequent changes in applying learning strategies are recommended.

A comprehensive experimental study could be done based on the intervention model (*Diagram 5.6.1*) in the future. Experimental studies are needed to test the hypotheses in the intervention model. It is a meaningful research question to be explored whether learning outcome could be improved after the two levels of interventions or not and to what extent compared with control groups. Future positive results based on the intervention model would be beneficial for teaching and learning, and even school context education reform.

The generalizability of the findings is limited by other factors as well. First, although the sample is representative of the school districts from which it was drawn, namely in China, it is not representative of adolescents across cultures. Hence, caution is warranted in generalizing the findings to adolescents of other backgrounds. Future research will need to explore cognition-emotions in learning mathematics in a variety of ethnic groups from a wide span of socioeconomic groups. Second, the current research relied on adolescents' self-reports, rather than behavioral observation, of their achievement goals and environmental factors within mathematics learning context. Adolescents may be biased in their reports, although self-report is a common approach (see Midgley et al., 1998; Schunk & Zimmerman, 1994). Self-report is the only way to capture the subjective contents of emotional experience. Self-report, however, has a few distinct disadvantages. Self-report cannot render real-time estimates of emotional experiences, and self-report measures are difficult to construct so that they render interval or ratio scales that capture more complex, non-linear relationships. In addition, self-report is not well suited to assess emotional processes that have limited access to consciousness (Pekrun, 2006). Third, the higher levels of negative emotions reported by girls in the study related to mathematics might have been due to higher levels of self-reported negative affectivity generally found in females rather than girls really experienced more negative affect than boys. Women are reported to experience emotions more intensely than men (e.g., Barrett, Robin, Pietromonaco, & Eyssell, 1998).

It was only a first step to look into adolescents' affective-cognitive processes in learning mathematics. There is a lot more to be done to understand their affective-cognitive processes in learning. Based on the current study gender was not one of the predictors for math achievement (*Table 4.5.5*). Could gender predict learners' math related career choices? And what other factors contribute to learners' career choices? What factors contribute substantially to their career orientations, cognitive ability, conation, and/or strategy development? What career orientations will high achievement females adopt compared with high achievement males? For these interesting questions longitudinal studies which catch changes and causal effects in natural settings over a relatively long time span are needed.

According to Schutz and DeCuir (2002), there are at least three approaches used by educational researchers: 1) investigating variables (e.g., looking at the relationship between variables); 2) investigating process and meaning; and 3) investigating social-historical contexts (e.g., how a variable is viewed differently in different cultures or time periods). Similar to most educational psychologists, the current inquiry on emotion, motivation, and cognition has been focused on variables. The other two approaches to inquire emotion, motivation, and cognition are necessary and needed in the future.

Bibliography

Abelson, R.P. (1983). Whatever became of consistency theory? *Personality and Social Psychology Bulletin, 9,* 37-54.
Abelson, R.P. (1985). A variance explanation paradox: When a little is a lot. *Psychological Bulletin, 97,* 129-133.
Ablard, K. E., & Lipschutz, R. E. (1998). Self-regulated learning in high-achieving students: Relations to advanced reasoning achievement goals, and sex. *Journal of Educational Psychology, 90,* 94-101.
Ackerman, P. L. (2006). Cognitive sex differences and mathematics and science achievement. American *Psychologist, 61,* 722-728.
Ainley, M. (2006). Connecting with learning: Motivation, affect, and cognition in interest process. *Educational Psychology Review, 18,* 391-405.
Ainsworth, M.D.S.(1973). The development of infant-mother attachment. In B. Caldwell & H.Ricciuti (Eds.), *Review of child development research* (Vol. 3, pp.173-196). Chicago: University of Chicago Press.
Alexander, P.A., Murphy, P.K., Woods, B.S., Duhon, K.E., & Parker, D (1997). College instruction and concomitant changes in students' knowledge, interest, and strategy use: A study of domain learning. *Contemporary Educational Psychology, 22,* 125-146.
Ames, C.(1990, April). Achievement goals and classroom structure: Developing a learning orientation in students. Paper presented at the annual meeting of the American Educational Research Association in Boston, MA.
Ames, C. (1992a). Achievement goals and the classroom motivational climate. In: D. H. Schunk & J. L. Meece (Eds), Student Perception in the Classroom (pp. 327-348). Hillsdale, NJ: Erlbaum.
Ames, C. (1992b). Classrooms: Goals, structures, and student motivation. *Journal of Educational Psychology, 84,* 261-271.
Ames, C., & Archer, J. (1988). Achievement goals in the classroom: Student learning strategies and motivation processes. *Journal of Educational Psychology, 80,* 260–267.
Anderman, L. (1999).Classroom goal orientation,school belonging, and social goals as predictors of students' positive and negative affect following the transition to middle school. *Journal of Research and Development in Education, 32,* 89-103.
Anderman, E. (2002). School effects on psychological outcomes during adolescence. *Journal of Educational Psychology, 94,* 795-809.
Anderman, E.M., & Maehr, M.L. (1994).Motivation and schooling in the middle grades. *Review of Educational Research, 64,* 287-309.
Anderman, E.M., & Midgley, C. (1997). Changes in achievement goal orientations, perceived academic competence, and grades across the transition to middle-level schools. *Contemporary Educational Psychology, 22,* 269-298.
Andre, T., Whigham, M., Hendrickson, A., & Chambers,S. (1999). Competency beliefs, positive affect, and gender stereotypes of elementary students and their parents about science versus other school subjects. *Journal of Research in Science Teaching, 36,* 719-747.
Andrews, G.R., & Debus, R.L. (1978).Persistence and causal perceptions of failure:Modifying cognitive attributions. *Journal of Educational Psychology, 70,* 154-166.
Arkinson, J.W. (1957). Motivational determinants of risk-taking behavior. *Psychological Review, 64,* 359-372.
Arkinson, J. W. (Ed.). (1958). *Motives in fantasy, action, and society.* Princeton, NJ: Van Nostrand.
Atkinson, J. W. (1964). *An introduction to motivation.* Princeton, NJ: Van Nostrand.
Arkinson, J.W. (1966). Motivational determinants of risk-taking behavior. In J. W. Atkinson and N.T. Feather (Eds.). *A theory of achievement motivation.* New York: Wiley.
Arkinson, J.W. (1974).The mainspring of achievement oriented activity. In J.W. Atkinson and J.O. Raynor (eds.), *Motivation and achievement.* Washington, DC: Winston.
Armstrong, J., & Kahl, S. (1978): *A national assessment of performance and participation of women in mathematics.* Prepared for National Institute of Education, Washington, DC.
Astleitner, H. (2000). Designing emotionally sound instruction. *Instructional Science, 28,* 169-198.

Atkinson, J. W. & Raynor, J. O. (1974). *Motivation and achievement.* Washington, DC: Winston.
Averill, J.A. (1983). Studies on anger and aggression. *American Psychologist, 38,* 1145-1160.
Baldwin, J.M. (1897). *Social and ethical interpretations in mental development.* New York: Macmillan.
Bandura, A. (1977). Self-efficacy:Toward a unifying theory of behavioral change. *Psychological Review, 84,* 191-215.
Bandura, A. (1986). *Social foundations of thought and action:A social cognitive theory.* Englewood Cliffs, NJ: Prentice-Hall.
Bandura, A. (1993). Perceived self-efficacy in cognitive development and functioning. *Educational Psychologist, 28,* 117-148.
Bandura, A. (1997). *Self-efficacy: The exercise of control.* New York: Freeman.
Bandura, A. (Ed.). (1995). *Self-efficacy in changing societies.* New York: Cambridge University Press.
Bargh, J. (1997). The automaticity of everyday life. In R. Wyer (Ed.), *Advances in social cognition* (Vol. 10, pp.1-61). Mahwah, NJ: LEA.
Barrett, L.F., Robin, L., Pietromonaco, P., &Eyssell, K. (1998).Are women the over emotional sex? Evidence from emotional experiences in social context. *Cognition and Emotion, 12,* 555-578.
Baumeister, R. (1993). Lying to yourself: The enigma of self-deception.In M. Lewis & C. Saarni (Eds.), *Lying and deception in everyday life* (pp. 166-183). New York: Guilford.
Bean, R.(1992). *The four conditions of self-esteem.* Santa Cruz., CA: ETR Associates.
Benjamin, M., McKeachie, W.J., Lin, Y. G., & Holinger, D. P. (1981).Test anxiety: Deficits in information processing. *Journal of Educational Psychology, 73,* 816–824.
Benjamin, M., McKeachie, W.J., & Lin, Y. G. (1987). Two types of est-anxious students: Support for an information processing model. *Journal of Educational Psychology, 79,* 131–136.
Bergin, D. A. (1999). Influences on classroom interest. *Educational Psychologist, 34,* 87-98. Csikszentmihalyi, M.(1997). *Finding flow: The psychology of engagement with everyday life.* New York: Basic Books.
Betancourt, H., & Weiner, B. (1982). Attributions for achievement-related events, expectancy, and sentiment. *Journal of Cross-Cultural Psychology, 13,* pp.362-374.
Biehler,R.,&Snowman,J.(1986).*Psychology Applied to Teaching* (5[th]ed.).Boston:Houghton Mifflin,1986.
Biggs, J.B. (1992). *Learning and schooling in ethnic Chinese:An Asian solution to a Western problem.* Unpublished manuscript, University of Hong Kong.
Bond, M. H. (1986). The Social Psychology of Chinese people. In M. H. Bond (Ed.), *The Psychology of the Chinese people* (pp. 213-264). Hong Kong: Oxford University Press.
Borkowski, J., Weyhing, R., & Carr, M. (1988).Effects of attributional retraining on strategy-based reading comprehension in learning-disabled students. *Journal of Educational Psychology, 80,*46-53.
Bottomley, D. (1990). Students strike trouble with "high aims." Published in *The South China Morning Post,* April 30, p.5.
Bowlby, J. (1969/1982). *Attachment and loss* (Vol. 1), 2nd ed. New York: Basic.
Bowlby, J. (1973). Attachment and loss (Vol. 2). New York: Basic.
Branden, N. (1993). *The six pillars of self-esteem.* New York: Bantam, Doubleday & Dell.
Brazelton, T.B. & Cramer, B. (1990). *The earliest relationship.* Reading,M.A: Addison-Wesley.
Breger, L. (1974). From instinct to identity. Englewood Cliffs, NJ: Prentice-Hall.
Brenner, E.M., & Salovey, P. (1997). Emotion regulation during childhood: developmental interpersonal, and individual considerations. In *Emotional Development and Emotional Intelligence,* Salovey, P., & Sluyter, D. J. (Eds.) Basic Books: A Division of Harper Collins Publishers.
Bretherton, I., Fritz, J., Zahn-Waxler, C., & Ridgeway, D. (1986). Learning to talk about emotions: A functionalist perspective. *Child Development, 57,* 529-548.
Briggs, D. (1970). *Your child's self-esteem.* Garden City, NY: Doubleday.
Bridges, L. J., & Grolnick, W.S. (1995). The development of emotional self regulation in infancy and early childhood. In N. Eisenberg (Ed.), *Review of Personality and Psychology* (pp. 185-211). Newbury Park, CA: Sage.
Bronfenbrenner,U.(1979).*The ecology of human development.*Cambridge,MA:Harvard University Press.
Brookover, W. B., & Erickson, E. L. (1975). Sociology of education. Homewood, IL: Dorsey. Brush, L. (1978). A validation study of the mathematics anxiety rating scale (MARS). *Educational and Psychological Measurement, 38,* 485-490.
Brophy, J. (1986). Teacher influences on student achievement. *American Psychologist, 41,* 1069-1077.

Brophy, J. (1999).Toward a model of the value aspects of motivation in education: Developing appreciation for particular learning domains and activities. *Educational Psychologist, 34,* 75-86.
Brown, A.L, Bransford, J.D., Campione, J.C.,&Ferrara, R. A. (1983).Learning, remembering and understanding. In J. Flavell & E. Markman (Eds.), *Handbook of child psychology: Vol. 3. Cognitive Development* (pp. 77–166). New York: Wiley.
Brown, J., & Weiner, B. (1984). Affective consequences of ability versus effort ascriptions: Controversies, resolutions, and quandaries. Journal of Educational Psychology, 76, 146-158.
Bruner, J. (1975). The ontogenesis of speech acts. *Journal of Child Language, 2,* 1-19.
Bruno, F. J. (1992). The family encyclopedia of child psychology and development. John Wiley & Sons, Inc.
Bull, B. A., & Drotar, D. (1991). Coping with cancer in remission: Stressors and strategies reported by children and adolescents. *Journal of Pediatric Psychology, 16,* 767-782.
Butcher, J. N. (1982). Cross-cultural research methods in clinical psychology. *Handbook of research methods in clinical psychology.* Edited by Philip C. Kendall & James N. Butcher. A Wiley-Interscience Publication.
Butcher, J. N., & Clarke, L. A. (1979). Recent trends in cross-cultural MMPI research and application. In J. N. Butcher (Ed.), *New developments in the use of the MMPI.* Minneapolis: University of Minnesota Press.
Butcher, J. N., & Pancheri, P. (1976): *Handbook of cross-national MMPI research.* Minneapolis: University of Minnesota Press.
Bye, D., Pushkar, D., & Conway, M. (2007). Motivation, interest, and positive affect in traditional and nontraditional undergraduate students. *Adult Education Quarterly, 57*(2), 141-158.
Cacioppo, J., Priester, J., & Berntson, G. (1993). Rudimentary determinants of attitudes: II. Arm flexion and extention have differential effects on attitudes. *Journal of Personality and Social Psychology, 65,* 5-17.
Campbell, D., Brislin, R., Stewart, V., & Werner, O. (1970). Back-translation and other translation techniques in cross-cultural research. Paper to be submitted to the International Journal of Psychology. In Brislin, R. W. (1970). Back-translation for cross-cultural research. *Journal of Cross-Cultural Psychology, 1,* 185-216.
Cantor, N., & Kihlstrom, J. (1987). *Personality and social intelligence.* Englewood Cliffs, NJ: Prentice-Hall.
Carnegie Council on Adolescent Development. (1989). *Turning points: Preparing American youth for the 21st century* (Report of the Task Force on Education of Young Adolescents). New York: Author.
Casey, M.B., Nuttall, R.L., & Pezaris, E. (1997).Mediators of gender differences in mathematics college entrance test scores: A comparison of spatial skills with internalized beliefs and anxieties. *Developmental Psychology, 33,* 669-680.
Caspi, A., Bem, D., &Elder, G.H., Jr.(1989).Continuities and consequences of interactional styles across the life course. *Journal of Personality, 57,* 375-406.
Chandler, T. A., Shama, D. D., & Wolf, F M. (1983). Gender differences in achievement and affiliation attributions: A five nation study. *Journal of Cross-Cultural Psychology, 14,* 241-256.
Chapin, M., & Dyck, D. G. (1976). Persistence in children's reading behavior as a function of N length and attribution retraining. *Journal of Abnormal Psychology, 85,* 511-515
Chen, C., & Stevenson, H. W. (1995). Culture and academic achievement: Ethnic and cross-national differences. *Advances in Motivation and Achievement, 9,* 119-151.
Church, M. A., Elliot, A. J., & Gable, S. L. (2001). Perceptions of classroom environment, achievement goals, and achievement outcomes. *Journal of Educational Psychology, 93,* 43-54.
Cohen, J. (1992). A power primer. *Psychological Bulletin, 112,* 155-159.
College Board. (2005). Summary reports: 2005. National report Retrieved January 8, 2008, from http://www.collegeboard.com/student/testing/ap/exgrd_sum/2005.html
Collins, B.E., Martin, J.C., Ashmore, R.D., & Ross, L. (1974). Some dimensions of the internal-external metaphor in theories of personality. *Journal of Personality and Social Psychology, 29,* 381-391.
Condry, J., & Dyer, S. (1976). Fear of success: Attribution of cause to the victim. *Journal of Social Issues, 33,* 63 – 83.

Cornell, D. G., Callahan, C. M., Bassin, L. E., & Ramsay, S. G. (1991). Affective Development in Accelerated Students. In *The academic acceleration of gifted children* by W. T. Southern, & E. D. Jones (Eds.). Teachers College Press.
Corno, L. (1986). The metacognitive control components of self-regulated learning. *Contemporary Educational Psychology, 11*, 333–346.
Corno, L., & Mandinach, E. (1983).The role of cognitive engagement in classroom learning and motivation. *Educational Psychologist, 18*, 88–100.
Corno, L., & Rohrkemper, M. (1985). The intrinsic motivation to learn in classrooms. In C. Ames & R. Ames (Eds.), *Research on motivation: Vol. 2. The classroom milieu* (pp. 53-90). New York: Academic Press.
Corno, L., & Snow, R. (1986). Adapting teaching to individual differences among learners. In M. Wittrock (Ed.), *Handbook of research on teaching* (pp. 605–629). New York: Macmillan. Nolen, S. (1988).Reasons for studying: Motivational orientations and study strategies. *Cognition and Instruction, 5*, 269–287.
Coopersmith, S. (1967). *The antecedents of self-esteem*. San Francisco: W. H. Freeman.
Covington, M. V. (1992). *Making the grade: A self-worth perspective on motivation and school reform.* Cambridge, MA: Cambridge University Press.
Covington, M.V., & Omelich, C.L (1984). An empirical examination of Weiner's critique of attribution research. *Journal of Educational Psychology, 76*, 1199-1213.
Crandall, V. C. (1969). Sex differences in expectancy of intellectual and academic reinforcement. In Smith, C.P. (Ed.). *Achievement-related motives in children*. Russell Sage Foundation, New York.
Culler, R.E., & Holahan, C.J.(1980).Test anxiety and academic performance:The effects of study related behaviors. *Journal of Educational Psychology, 72*, 16–20.
Dai, Y.D. (2006). There is more to aptitude than cognitive capacities. *American Psychologist, 61,7*, 723-724.
Dai, Y.D., Moon, S.M., & Feldhusen, J.M.(1998). Achievement motivation and gifted students: A social cognitive perspective. *Educational Psychologist, 33*, 45-63.
Damasio, A. R. (1994). Descartes' error: Emotion, reason, and the human brain. New York: Grosset / Putnam.
Davidson, R. J., Scherer, K. R., & Goldsmith, H. H. (2003).(Eds.). *Handbook of affective sciences.* New York: Oxford University Press.
Deci, E. L., & Ryan, R. M. (1985). *Intrinsic motivation and self-determination in human behavior.* New York: Plenum Press.
Deci, E. L., & Ryan, R. M. (2000). The "what" and "why" of goal pursuits: Human needs and self-determination of behavior. *Psychological Inquiry, 11*, 227-268.
De Vos, G. A. (1968). Achievement and innovation in culture and personality. In E. Norbeck, D. Price-Williams, & W. M. McCord (Eds.), *The Study of Personality. An interdisciplinary approach* (pp. 348-370). New York: Holt, Rinchart & Winston.
Defalco, K. (1997). Educator's comment. In *Emotional development and emotional intelligence:Educational implications.* Salovey, P., & Sluyter, D. J. (Eds.). Basic Books: A Division of Harper Collins Publishers.
Dodge, K.A. (1989). Coordinating responses to aversive stimuli: Introduction to a special section on the development of emotion regulation. *Developmental Psychology, 25*, 339-342.
Dupont,H.(1994).*Emotional development, theory and applications:A Neo-Piagetian perspective.* Praeger.
Davies, P. G., Spencer, S. J., Quinn, D. M., & Gerhardstein, R. (2002). Consuming images: How television commercials that elicit stereotype threat can restrain women academically and professionally. *Personality and Social Psychology Bulletin, 28*, 1615-1628.
Dweck, C. (1975). The role of expectations and attributions in the alleviation of learned helplessness. *Journal of Personality and Social Psychology, 31*, 674-685.
Dweck, C. (1986). Motivational processes affecting learning. *American Psychologist, 41*, 1040-1048.
Dweck, C. (1999). *Self-theories: Their role in motivation, personality, and development.* Philadelphia: Psychology Press.
Dweck, C., & Elliott, E. (1983). Achievement motivation. In E. M. Heatherington (Ed.), *Handbook of child psychology:Vol.4.Socialization, personality, and social development.*(pp. 643-691).New York: Wiley.

Dweck, C. S., & Goetz, T.E. (1978). Attributions and earned helplessness. In Harvey, J. H., Ickes, W., & Kidd, R. F. (Eds.). *New directions in attribution research. Vol. 2.* Hillsdale, NJ: Lawrence Erlbaum.

Dweck, C. S., & Leggett, E. L. (1988). A social-cognitive approach to motivation and personality. *Psychological Review, 95,* 256-273.

Dweck, C.S., Mangels, J.A., & Good, C. (2004).Motivational effects on attention, cognition, and performance. In D.Y. Dai. & R.J. Sternberg (Eds.), *Motivation, emotion, and cognition:Integrative perspectives on intellectual functioning and development* (pp. 41-55). Mahwah, NJ: Erlbaum.

Early, P.C. (1993). East meets West meets Midwest: Further explorations of collectivistic and individualistic work groups. *Academy of Management Journal, 36,* 319-348.

Eaton, M.J., & Dembo, M.H. (1997).Differences in the motivational beliefs of Asian American and non-Asian students. *Journal of Educational Psychology, 89,* 433-440.

Eccles, J. (1983). Expectancies, values and academic behaviors. In J.T. Spence (Ed.), *Achievement and achievement motives* (pp. 75–146). San Francisco: Freeman.

Eccles, J.S., Adler, T., & Meece, J.L. (1984). Sex differences in achievement: A test of alternate theories. *Journal of Personality and Social Psychology, 46,* 26-43.

Eccles, J.S., & Jacob, J.E.(1986). Social forces shape math attitudes and performance. *Signs: Journal of Women in Culture and Society, 11,* 367-380.

Eccles, J.S., Jacob, J.E., & Harold, R.E. (1990). Gender role stereotypes, expectancy effects, and parents' socialization of gender differences. *Journal of Social Issues, 46,* 183-201.

Eccles, J.S., & Midgley, C. (1989). Stage/environment fit: Developmentally appropriate classrooms for early adolescents. In R. E. Ames & C. Ames (Eds.), *Research on motivation in education* (Vol. 3, 139-186). New York: Academic.

Eccles, J.S., Midgley, C., Wigfield, A., Miller-Buchannan, C., Reuman, D., Flanagan, C., & Maclver, D. (1993b). Development during adolescence: The impact of stage-environment fit on young adolescents' experiences in schools and families. *American psychologist, 48,* 90-101.

Eccles, J.S., Wigfield, A., Harold, R.D., Blumenfeld, P. (1993c).Ontogeny of children's self-perceptions and subjective task values across activity domains during the early elementary school years. *Child Development, 64,* 830-847.

Eccles, J.S., Wigfield, A., Midgley, C., Reuman, D., MacIver, D., & Feldlaufer, H. (1993a). Negative effects of traditional middle schools on students'motivation.*Elementary School Journal,93,*553-574.

Edelman, G. (1987). *Neurodarwinism.* New York: Basic.

Egeland, B., Pianta, R., & O'Brine, M. (1993).Maternal intrusiveness in infancy and child maladaptation in the early school years. *Development and Psychopathology, 81,* 359-370.

Eid, M., & Diener, E. (2001). Norms for experiencing emotions in different cultures: Inter- and intranational differences. *Journal of Personality and Social Psychology, 81,* 869-885.

Eisenberg, N., & Fabes, R. A. (1992). Emotion, regulation, and the development of social competence. In M. S. Clark (Ed.), *Review of personality and social psychology: Emotion and social behavior* (Vol. 14, pp.119-150). Newbury Park, CA: Sage.

Elliot, A. (1997). Integrating the "classic" and "contemporary" approaches to achievement motivation: A hierarchical model of approach and avoidance achievement motivation. In M.Maehr & P.Pintrich (Eds.), *Advances in motivation and achievement* (pp.143-179). Greenwich, CT: JAI.

Elliot, A. (1999). Approach and avoidance motivation and achievement goals. *Educational Psychologist, 34,* 169-189.

Elliot, A., & Church, M. (1997). A hierarchical model of approach and avoidance achievement motivation. *Journal of Personality and Social Psychology, 72,* 218-232.

Elliot, A., & Covington, M. (in press). Approach and avoidance motivation. *Educational Psychology Review.*

Elliott, E.S., & Dweck, C.S. (1988). Goals: An approach to motivation and achievement. *Journal of Personality and Social Psychology, 54,* 5-12.

Elliot, A., & Harackiewicz, J. (1996). Approach and avoidance achievement goals and intrinsic motivation: A mediational analysis. *Journal of Personality and Social Psychology, 70,* 461-475.

Elliot, A.J., & McGregor, G.A. (1999). Test anxiety and hierarchical model of approach and avoidance achievement motivation. *Journal of Personality and Social Psychology, 76,* 628-644.

Elliot, A.J., & McGregor, G.A. (2001). A 2 x 2 Achievement Goal Framework. *Journal of Personality and Social Psychology, 80,* 501-519.

Ember, C., & Levinson, D. (1991). The substantive contributions of world-wide cross-cultural studies using secondary data. *Behavioral Science Research, 25*, 79-140.
Epstein, J.L.(1989).Family structures and student motivation: A developmental perspective. In: C. Ames & R. Ames (Eds), *Research on Motivation in Education* (Vol. 3, pp. 259-295). San Diego, CA: Academic Press.
Erikson, E. H. (1959). Identity and the life cycle. *Psychology, 1.*
Ernest, J. (1976). *Mathematics and sex.* Santa Barbara: University of California Press.
Eynde, P.O., & Turner, J.E. (2006). Focusing on the complexity of emotion issues in academic learning: A dynamical component systems approach. *Educational Psychology Review, 18*, 361-376.
Fairbank, J.K., & Reischauer, E. O. (1973). *China: Tradition and Transformation.* New York: Houghton Mifflin.
Farmer, H.S., Wardrop, J.L., Anderson, M.Z., & Risinger, R. (1995). Women's career choices: Focus on science, math and technology careers. *Journal of Counseling Psychology, 42*, 155-170.
Fehrenbach, C. R. (1991). Gifted/average readers: Do they use the same reading strategies? *Gifted Child Quarterly, 35*, 125-127.
Feldhusen, J.F., & Dai, D.Y. (1997).Gifted students' attitudes and perceptions of the gifted label, special programs, and peer relations. *Journal of Secondary Gifted Education, 9*, 15-20.
Fennema, E., & Sherman, J. (1977). Sex-related differences in mathematics achievement, spatial visualization and affective factors. *American Educational Research Journal, 14*, 51-71.
Fincham, F., & Cain, K. (1986). Learned helplessness in humans: A developmental analysis. *Developmental Review, 6*, 25-86.
Flett, G., Hewitt, P., Blankstein, K., & Gray, L. (1998). Psychological distress and the frequency of perfectionist thinking. *Journal of Personality and Social Psychology, 75*, 1363-1381.
Fogel, A. (1982). Affect dynamics in early infancy: Affective tolerance. In T. Field & A. Fogel (Eds.), *Emotion and early interaction.* Hillsdale, NJ: Erlbaum.
Fogel, A. (1993). *eveloping through relationships:Origins of communication, self, and culture.* Chicago, IL: University of Chicago Press.
Fogel, A., & Thelen, E. (1987). Development of early expressive and communicative action: Reinterpreting the evidence from a dynamic systems perspective. *Developmental Psychology, 23*, 747-761.
Forster, J., Higgins, E., & Idson, L. (1998).Approach and avoidance strength during goal attainment: Regulatory focus and the "goal looms larger" effect. *Journal of Personality and Social Psychology, 75*, 1115-1131.
Fouad, N. A. (1993). Cross-cultural vocational assessment. *The Career Development Quarterly, 42*,4-13.
Fox, L. (1977). The effects of sex-role socialization on mathematics participation and achievement. *Women and mathematics:Research perspectives for change.* NIE Papers in Education and Work, No. 8.
Fraser, B.J.(1994).Research on classroom and school climate.In D.L. Gabel (Ed.), *Handbook of research on science teaching and learning* (pp.493-541). New York: Macmillan.
Frasier, B., & Fisher, D. (1986). Using short forms of classroom climate instruments to assess and improve classroom psychosocial environment. *Journal of Research in Science Teaching, 23*, 387-413.
Fredrickson, B. L., & Losada, M. F. (2005). Positive affect and the complex dynamics of human flourishing. *American Psychologist, 60*, 678-686.
Frenzel, A.C., Pekrun, R., & Goetz, T. (2007a). Perceived learning environment and students' emotional experiences: A multilevel analysis of mathematics classrooms.*Learning and Instruction,17*,478-493.
Frenzel, A.C., Pekrun, R., & Goetz, T. (2007b). Girls and mathematics – A "hopeless" issue? A control-value approach to gender differences in emotions towards mathematics. *European Journal of Psychology of Education, 22*, 497-514.
Frenzel, A.C., Pekrun, R., Goetz, T., & vom Hofe, R. (2006). *Girls' and boys' emotional experiences in mathematics.* Manuscript submitted for publication.
Frenzel, A.C., Thrash, T.M., Pekrun, R., & Goetz, T. (2007c). Achievement emotions in Germany and China:A cross-cultural validation of the Academic Emotions Questionnaire – Mathematics. *Journal of Cross-Cultural Psychology, 38*, 302-309. Retrieved at UB Muenchen on February 18, 2008 from http://jcc.sagepub.com/cgi/content/abstract/38/3/302
Frieze, I. H., Fisher, J., Hanusa, B., McHugh, M., & Valle, V. (1978). Attributing the causes of success and failure: Internal and external barriers to achievement in women. In Sherman, J., & Denmark, F. (Eds.). *Psychology of women: Future directions of research.* New York: Psychological Dimensions.

Frijda, N.H. (1988). The laws of emotion. *American Psychologist, 43,* 349-358.
Frome, P.M., & Eccles, J.S. (1998). Parents' influence on children's achievement-related perceptions. *Journal of Personality and Social Psychology, 74,* 435-452.
Fry, P.S., & Ghosh, R. (1980). Attribution of success and failure. *Journal of Cross-cultural Psychology, 11*(3), 343-363.
Gallimore, R., Boggs, J.W., & Jordan, C. (1974). *Culture, behavior, and education:A study of Hawaian-Americans.*Beverly Hills: Sage.
Garber, J., Quiggle, N.L., Panak, W., & Dodge, K.A. (1991). Aggression and depression in children:Comorbidity, specificity, and social cognitive processing. In D. Cicchetti & S.L. Toth (Eds.), *Internalizing and externalizing expressions of dysfunction:Rochester symposium on developmental psychopathology* (Vol. 2). Hillsdale, NJ: Erlbaum.
Garcia, T., & Pintrich, P.R. (1996). Assessing students' motivation and learning strategies in the classroom context: the Motivated Strategies for Learning Questionnaire. In M. Birenbaum and F.J.R.C. Dochy (Eds.), *Alternatives in assessment of achievements, learning processes and prior knowledge* (pp. 319-339). Boston: Kluwer Academic.
Garner, R., & Alexander, P. (1989). Metacognition: Answered and unanswered questions. *Educational Psychologist, 24,* 143–158.
Geary, D. (1995). Reflections of evolution and culture in children's cognition. *American Psychologist, 50,* 24-37.
Gebelt, J.L. (1995). *Identity, emotion and memory in college students.* Unpublished doctoral dissertation, Rutgers – The State University of New Jersey, New Brunswick, NJ.
Geertz, C. (1975). On the nature of anthropological understanding. *American Scientist, 63,* 47-53.
Gerrig, R. J., & Zimbardo, P. G. (2002). *Psychology and life.* Boston: Allyn & Bacon.
Gibson, E. J. (1982). The concept of affordance in development: The renascence of functionalism. In W. A. Collins (Ed.), *The concept of development. Minnesota symposium on child psychology* (Vol. 15, pp. 55-81).
Glick, J. (1992). Werner's relevance for contemporary developmental psychology. *Developmental Psychology, 28,* 558-565.
Goetz, T. (2004). Emotionales Erleben und selbstreguliertes Lernen bei Schülern im Fach Mathematik (Students emotions and self-regulated learning in mathematics). Munich, Germany: Utz.
Goetz, T., Frenzel, A.,C., Pekrun,R., & Hall, N.C. (2006).The domain specificity of academic emotional experiences. *The Journal of Experimental Education, 75,* 5-29.
Goetz, T., Pekrun, R., Hall, N., & Haag, L. (2006). Academic emotions from a social-cognitive perspective:antecedents and domain specificity of students'affect in the context of Latin instruction. *British Journal of Educational Psychology, 76*(2), 289-308.
Gottfried, A.E., & Gottfried, A.W.(1996). A longitudinal study of academic intrinsic motivation in intellectually gifted children:Childhood through early adolescence.*Gifted Child Quarterly,40,*179-183.
Gottlieb, G. (1991). Experiential canalization of behavioral development: Theory. *Developmental Psychology, 27,* 4-13.
Grant, H., & Dweck, C.S. (2003). Clarifying achievement goals and their impact. *Journal of Personality and Social Psychology, 85,* 541-553.
Greenberg, M.T., & Snell, J.L. (1997).Brain development and emotional development:The role of teaching in organizing the frontal lobe. In *Emotional development and emotional intelligence.* Salovey,P., & Sluyter, D. J. (Eds.). Basic Books: A Division of Harper Collins Publishers.
Greenwald, A. G., & Pratkanis, A. R. (1984). The self. In R. S. Wyer & T. K. Srull (Eds.), Handbook of social cognition (Vol. 3, pp. 129-178). Hillsdale, NJ: Erlbaum.
Gross, J. J. (1998). The emerging field of emotion regulation: An integrative review. *Review of General Psychology, 2,* 271-299.
Gudykunst, W.B., Matsumoto, Y., Ting-Tommey, S., Nishida, T., Kim, K.S., & Heyman, S. (1996). The influence of cultural individualism-collectivism, self-construals, and values on communication styles across cultures. *Human Communication Research, 22,* 510-543.
Gysbers, N.C., Heppner, M.J., & Johnston, J.A. (1998). *Career counseling:Process, issues, and techniques.* MA:Allyn & Bacon.
Hall, G. S. (1904). *Adolescence: Its psychology and its relations to physiology, anthropology, sociology, sex, crime, religion, and education* (Vols. 1,2). New York: Appleton.

Halpern, D.(2000).Sex differences in achievement scores, can we design assessments that are fair, meaningful, and valid for girls and boys? *Issues in education, 8,* 1-19.

Harter, S. (1980). A model of intrinsic mastery motivation in children: Individual differences and developmental change. *Minnesota Symposium on Child Psychology. Vol. 14.* Hillsdale, NJ: Lawrence Erlbaum.

Harter, S. (1981).The new self-report scale of intrinsic versus extrinsic orientation in the classroom: Motivational and informational components. *Developmental Psychology, 17,* 300-312.

Harter, S. (1986). Processes underlying the construction, maintenance and enhancement of the self-concept in children. In J. Suls & A. Greenwald (Eds.), *Psychological Perspectives on the Self.*(Vol. 3, 136-182). Hillsdale, NJ: Erlbaum.

Harackiewicz, J.M., & Elliot, A.J. (1998). The joint effects of target and purpose goals on intrinsic motivation: A mediational analysis. *Personality and Social Psychology Bulletin, 24,* 675-689.

Harvey, J.H., & Weary, G. (1981). *Perspectives on attributional processes.* Dubuque, IA: Wm.C. Brown.

Haviland-Jones, J., Gebelt, J. L., & Stapley, J. C. (1997). The questions of development in emotion. In *Emotional development and emotional intelligence.*

Hayduk, L. A. (1987). *Structural equation modeling with LISREL: Essentials and advances.* Baltimore: Johns Hopkins University Press.

He, S. J. (2004). Emotions in learning mathematics in China and Germany – A cross-cultural study. Unpublished master thesis. University of Munich, Munich, Germany.

Heelas, P. (1986).Emotion talk across cultures. In R.M. Harré (Ed.), *The social construction of emotions* (pp.234-266). Oxford, England: Basil Blackwell.

Heider, F. (1958). *The psychology of interpersonal relations.* New York: Wiley.

Heller, K., Futterman, R., Kaczala, C., Karabenick, J.D.,& Parsons, J (1978). *Expectancies, utility values, and attributions for performance in mathematics.* Paper presented at meeting of American Educational Research Association, Toronto.

Hess, R.D., Chang, C.M.,& McDevitt, T.M. (1987).Cultural variations in family beliefs about children's performance in Mathematics: Comparisons among People's Republic of China, Chinese American, and Caucasian-American Families. *Journal of Educational Psychology, 79(2),*179-188.

Higgins, E. T., & Trope,Y. (1990). Activity engagement theory: Implications of multiple identifiable input for intrinsic motivation. In E.T. Higgines & R. M. Sorrentino (Eds.), *Handbook of motivation and cognition: Foundations of social behavior; Vol. 2* (pp.229-264). New York: Guilford.

Hill, J. P: (1980). *Understanding early adolescence: A framework.* Chapel Hill, NC: Center for Early Adolescence.

Hill, K., & Wigfield, A. (1984). Test anxiety: A major educational problem and what can be done about it. *Elementary School Journal, 85,* 105–126.

Ho, D.Y.F. (1986). Chinese Patterns of Socialization: A Critical Review. In M.H. Bond (Ed.), *The Psychology of Chinese people* (pp. 1-35). Hong Kong: Oxford University Press.

Hofer, B. K., & Yu, S. L. (2003). Teaching self-regulated learning through a "learning to learn" course. *Teaching of Psychology, 30,* 30-33.

Hoffman, M. L. (1982).Development of prosocial motivation: Empathy and guilt. In N. Eisenberg-Borg (Ed.), *Development of prosocial behavior* (pp. 281-313). New York: Academic Press.

Hofstede, G. (1983). Dimensions of national cultures in fifty countries and three regions. In J.B. Deregowski, S.Dziurawiec, and R.C. Annis (Eds.), *Expiscations in Cross-cultural Psychology* (pp. 335-355). Lisse, Netherlands: Swets and Zeitlinger. Expectations

Holloway, S.D. (1988). Concepts of ability and effort in Japan and the United States. *Review of Educational Research, 58,* 327-343.

Hong, Y., Chiu, C., Dweck, C. S., Lin, D. M., & Wan, W. (1999). Implicit theories, attributions, and coping: A meaning system approach. *Journal of Personality and Social Psychology, 77, 588-599.*

Horner, M. (1968). Sex differences in achievement motivation and performance in competitive and non-competitive situations. Unpublished doctoral dissertation, University of Michigan.

Horner, M. S. (1972). Toward an understanding of achievement-related conflicts in women. *J. soc., 28,* 147-176.

Hsu, F.L.K. (1967). *Under the Ancestor's Shadow: Kinship, Personality and Social Mobility in Villane China* (revised and expanded edition). New York: Doubleday.

Hsu, F.L.K.(1985).The Self in Cross-Cultural Context.In A.J.Marsetta, G. De Vos,& F. L. K. Hsu (Eds.), *Culture and Self*(pp. 24-55). London: Tavstock.

Hsueh, W. (1997). *A cross-cultural comparison of gifted children's theories of intelligence, goal orientations, and response to challenge*. Unpublished doctoral dissertation. Purdue University, West Lafayette, IN.

Hua, K., & Salili, F. (1989). Attribution of examination results – Chinese primary school students in Hong Kong. *Psychologia, 32,* 163-171.

Hua, K., & Salili, F. (1990). Examination result attribution, expectancy, and achievement goals among Chinese students in Hong Kong. *Educational Studies, 16*(1), 17-31.

Hua, K., & Salili, F. (1991). Structure and semantic differential placement of specific causes: Academic causal attributions by Chinese students in Hong Kong. International *Journal of Psychology, 26*(2), 175-193.

Hubbart,J., & Coie, J.D.(1994).Emotional determinants of social competence in children's peer relationships. *Merrill-Palmer Quarterly, 40,* 1-20.

Hui,C.H(1988)Measurement of Individualism-Collectivism.*Journal of Research in Personality,22,*17-36.

Hwang, K. K. (1990). Modernization of the Chinese Family Business. *International Journal of Psychology, 25,* 593-618.

Hyde, J.S., Fennema, E., Ryan, M., Frost, L.A., & Hopp, C. (1990). Gender comparisons of mathematics attitudes and affect: A meta-analysis. *Psychology of Women Quarterly, 14,* 299-324.

Iben, M. F. (1991). Attitudes and mathematics. *Comparative Education, 27,* 135-151.

Jackson, D. N., Ahmed, S. A., & Heapy, N. A. (1976). Is achievement motivation a unitary construct? *Journal of Research in Personality, 10,* 1-21.

Jacob, B. (1996). Leistungsemotionen by Schülern (Achievement emotions in school students). Unpublished master's thesis, University of Regensburg, Germany.

Jacobs, J. E., & Eccles, J. S. (1992). The impact of mothers'gender-role stereotypic beliefs on mothers' and children's ability perceptions. *Journal of Personality and Social Psychology, 63,* 932-944.

Jacobs, J.E., Finken, L.L., Griffin, N.L., & Wright, J.D. (1998).The career plans of science-talented rural adolescent girls. *American Education Research Journal, 35,* 681-704.

Jacobs, W.J., & Nadel,L. (1985). Stress-induced recovery of fears and phobias. *Psychological Review, 92,* 512-531.

Jacobs, J. E., Lanza, S., Osgood, D.W., Eccles, J.S., & Wigfield, A. (2002). Changes in children's self-competence and values: Gender and domain differences across grades one through twelve. *Child Development, 73,* 509-527.

Jagacinski, C. M., & Nicholls, J. G. (1984). Conception of ability and related affects in task involvement and ego involvement. *Journal of Educational Psychology, 76,* 909-919.

James, W. (1890). *Principles of psychology*. New York: Holt, Rinehart and Winston.

Jenkins-Friedman, R. & Murphy, D.L. (1988). The Mary Poppins effect: Relations between gifted students' self concept and adjustment. *Roeper Review,* 11(1), 26-30.

Jones, C. J. (1992). Social and emotional development of exceptional students: handicapped and gifted. Charles C Thomas Publisher.

Jones, C.J. (1992b).Enhancing Self-Concepts and Achievement of Mildly Handicapped Students: Learning Disabled, Mildly Mentally Retarded, Behavior Disordered, and Speech/Language Impaired. Springfield, IL: Charles C Thomas.

Joreskog, K.G., & Sorbom, D. (1986). *LISREL: Analysis of linear structural relationships by the method of maximum likelihood: User's guide*. Mooresville, IN: Scientific Software Inc.

Kagan, J., & Snidman, N. (1991).Temperamental factors in human development. *American Psychologist, 46,* 856-862.

Kaminski, D., Erickson, E., Ross, M, & Bradfield, L. (1976). *Why females don't like mathematics.* The effect of parental expectations. Paper presented at meeting of American Sociological Association, New York.

Kaplan, A., & Maehr, M. L. (1999). Achievement goals and student well-being. *Contemporary Educational Psychology, 24,* 330-358.

Kaplan, A., & Midgley, C. (1999). The relationship between perceptions of the classroom goal structure and early adolescents'affect in school: The mediating role of coping strategies. *Learning and Individual Differences, 11,* 187-212.

Kashdan, T. B., & Fincham, F. D. (2004). Facilitating curiosity: A social and self-regulatory perspective for scientically based interventions.In P.A. Linley & S. Joseph (Eds.). *Positive psychology in practice* (pp. 482-503). Hoboken, NJ: John Wiley.

Kazdin, A. E., & Bootzin, R. R. (1972). The token economy: An evaluative review. *Journal of Applied Behavior Analysis, 5,* 343 – 372.

Kellerman, H. (1983). An epigenetic theory of emotions in early development. In R. Plutchik & H. Kellerman (Eds.), *Emotion: Theory, research, and experience* (pp. 315-349). New York: Academic.

Kelley,H.H.(1971).*Attributions in social interactions.* Morristown, NJ: General Learning Press. Kenney-Benson, G.A., Pomerantz, E.M., Ryan, A.M., & Patrick, H. (2006). Sex differences in math performance: The role of children's approach to schoolwork. *Developmental Psychology, 42,* 11-26.

Kenniston,K. (1965).Social change and youth in America. In E.H.Erikson, (Ed.). *The challenge of youth.,* N.Y.: Anchor.

Kenny, D.A., Kashy, D.A., & Bolger, N. (1998). Data analysis in social psychology. In D. T. Gilbert, S. T.Fiske, & G.Lindzey (Eds.), *The handbook of social psychology* (4th ed., Vol.1, pp. 233-265).New York: McGraw-Hill.

Kerr, B., Colangelo, N., & Gaeth, J. (1988). Gifted adolescents' attitudes toward their giftedness. *Gifted Child Quarterly,* 32(2), 245-247.

Knollmann, M., & Wild, Elke. (2007). Quality of parental support and students' emotions during homework: Moderating effects of students' motivational orientations. *European Journal of Psychology of Education, 22,* 63-76.

LaFreniere, P. J. (2000). *Emotional development: A biosocial perspective.* Wadsworth: Thomson Learning.

Lamb, M.E. (1981). The development of social expectations in the first year of life. In M.-e. Lamb & L. R., Sherwood (Eds.). *Infant social cognition: Empirical and theoretical consequences* (pp. 155-175). Hillsdale, NJ: Erlbaum.

Lau, S. (1986). The value orientation and education process of Chinese University Students in Hong Kong. *Educational Journal, 14*(2), 7-13.

Lau, Kit-Ling., & Lee, John C. K.(2008).Validation of a Chinese achievement goal orientation questionnaire. *British Journal of Educational Psychology, 78,* 331-353.

Lazarus, R. (1991). *Emotion and adaptation.* New York: Oxford University Press.

LaFreniere, P. J. (2000). *Emotional development: A biosocial perspective.* Wadsworth: Thomson Learning.

Lazarus, R. S., & Folkman, S. (1984). *Stress, appraisal, and coping.* New York: Springer.

LeDoux, J. E. (1995). Emotion: Clues from the brain. *Annual Review of Psychology, 46,* 209-235.

Lent, R.W., Brown, S.D., & Gore, P.A. (1997). Discriminant and predictive validity of academic self-concept, academic self-efficacy, and mathematics-specific self-efficacy. *Journal of Counseling Psychology, 44,* 307-315.

Levine, J. (1989). *Secondary Instruction: A Manual for Classroom Teaching.* Boston: Allyn & Bacon.

Lewis, M., & Michalson, L. (1982). The socialization of emotions. In T. Field & A.Fogel (Eds.), *Emotion and early interaction* (pp.189-212). Hillsdale, NJ: Erlbaum.

Lewis, M., & Michalson, L.(1983).*Children's emotions and moods:Developmental theory and measurement.* New York: Plenum.

Linnenbrink, E.A.(2002).*The dilemma of performance goals:Promoting students' motivation and Learning in cooperative groups.* Unpublished dissertation. The University of Michigan, Ann Arbor, MI.

Linnenbrink, E.A. (2004). Person and context: Theoretical and practical concerns in achievement goal theory. *Motivating students, improving schools. Advances in Motivation and Achievement, 13,* 159-184. Elsevier Ltd.

Linnenbrink, E.A., & Pintrich, P.R. (2002). Achievement goal theory and affect: An asymmetrical bidirectional model. *Educational Psychologist, 37,* 69-78.

Lofgreen, K.B., & Larson, A. (1992). Key components of self esteem. In Mönks, F, & Peters, W. (Eds.) (1992). *Talent for the future.* Van Gorcum, Assen/Maastricht, The Netherlands.

Lynn, R., & Irwing, P. (2005). Sex differences in means and variability on the progressive matrices in universitiy students: A meta-analysis. *British Journal of Psychology, 96,* 505-524.

Lutz, C. (1988). *Unnatural emotions: Everyday sentiments on a Micronesian atoll and their challenge to Western theory.* Chicago: University of Chicago Press.

Madison, P. (1969). *Personality development in college.* Reading, Mass.: Addison-Wesley.
Maehr, M.L. (1978). Sociocultural origins of achievement motivation. In D.Bar-Tal, and L. Sax (Eds.), *Social Psychology of education: Theory and research.* New York: John Wiley & Son.
Maehr, M. (1989). Thoughts about motivation. In C. Ames & R. Ames (Eds.), *Research on motivation in education* (Vol. 3, pp.299-315). New York: Academic Press.
MacIver, D.J., & Epstein, J.L.(1993).Middle grades research: Not yet mature, but no longer a child. *Elementary School Journal, 93, 519-531.*
Maehr, M.L. (1991).The "psychological environment" of the school: A focus for school leadership. In P. Thurston & P. Zodhiates (Eds.), *Advances in educational administration* (pp. 51-81). Greenwich, CT: JAI.
Maehr, M.L., & Anderman, E.M. (1993). Reinventing schools for early adolescents: Emphasizing task goals. *Elementary School Journal, 93,* 593-610.
Maehr, M.L., & Midgley, C. (1991).Enhancing student motivation: A schoolwide approach. *Educational Psychologist, 26,* 399-427.
Maehr, M.L., Medgley, C., & Urdan, T. (1992). School leader as motivator. *Educational Administration Quarterly, 18,* 412-431.
Maehr, M.L., & Pintrich, P.R. (Eds.). (1991). *Advances in motivation and achievement: Goals and self-regulatory processes (Vol. 7).* Greenwich, CT: JAI.
Maehr, M.L, Pintrich, P.R., &Zimmerman, M.(1993).*Personal and contextual influences on adolescent wellness.* Unpublished manuscript.
Magnussun, D.(1988).*Individual development from an interactional perspective.* Hillsdale, NJ: Erlbaum.
Mahe, M.L. (1984). Meaning and motivation: Toward a theory of personal investment. In: R. Ames & C. Ames (Eds.), *Research on Motivation in Education: Student Motivation* (Vol. 1, pp. 115-143). New York: Academic Press.
Manaster, G.J.(1989).*Adolescent Development: A Psychological Interpretation.* Itasca, IL: F.E. Peacock.
Mandler, G., & Sarason, S.B. (1952). A study of anxiety and learning. *Journal of Abnormal and Social Psychology, 47,* 166-173.
Markus, H.R., & Kitayama, S. (1991). Culture and the self: Implication for cognition, emotion, and motivation. *Psychological Review, 98,* 224-253.
Markus, H., & Wurf, E. (1987). The dynamic self-concept: A social psychological perspective. *Annual Review of Psychology: 38,* 299-337.
Marlowe, W.B. (1992).The impact of a right prefrontal lesion on the developing brain. Brain and Cognition, 20, 205-213.
Marsh, H. W. (1989). Age and sex effects in multiple dimensions of self-concept: Preadolescence to early childhood. *Journal of Educational Psychology, 81,* 417-430.
Marsh, H. W., & Craven, R.(1997). Academic self-concept: Beyond the Dustbowl. In G. D. Phye (Ed.), *Handbook of classroom assessment:Learning, achievement and adjustment* (pp. 131-198). Orlando, FL: Academic Press.
Martell, R.F., Lane, D.M., & Emrich, C.E. (1996). Male-female differences: A computer situation. *American Psychologist, 51,* 157-158.
Matthews, G., Zeidner, M., & Roberts, R.D. (2002). *Emotional intelligence: Science and myth.* Cambridge, MA: MIT Press.
Mayer, J.D., & Salovey, P. (1997). What is emotional intelligence? In *Emotional development and emotional intelligence: Educational implications.* Salovey, P., & Sluyter, D. J.(Eds.). Basic Books: A Division of Harper Collins Publishers.
McRobbie, C.J., & Fraser, B.J. (1993). Associations between student outcomes and psychosocial science environment. *Journal of Educational Research, 87,* 78-85.
Meece, J. (1991). The classroom context and children's motivational goals. In M. Maehr & P. Pintrich (Eds.), *Advances in achievement motivation research* (pp. 261-286). Greenwich, CT: JAI.
Meece, J., Blumenfeld, P., & Hoyle, R. (1988). Students' goal orientations and cognitive engagement in classroom activities. *Journal of Educational Psychology, 80,* 514–523.
Meece, J.L., & Holt, K.(1993).A pattern analysis of students'achievement goals. *Journal of Educational Psychology, 85,* 582-590.
Mertler, C., & Vannatta, R. (2001). *Advanced and multivariate statistical methods:Practical application and interpretation.* Los Angeles:

Mesquita, B., & Frijda, N.H. (1992). Cultural variations in emotions: A review. *Psychological Bulletin, 112,* 179-204.
McClelland, D. C. (1961). *The achieving society.* New York: Free Press.
McClelland, D. C., Atkinson, J., Clark, R., & Lowell, E. (1953).*The achievement motivation.* Glenview, IL: Scott, Foresman.
McClelland, D. C., Atkinson, J.W., Clark, R.A., & Lowell, E. L. (1953). *The achievement motive.* New York: Appleton-Century-Crofts.
Mead, G. H. (1934). *Mind, self, and society.* Chicago: University of Chicago Press.
Middleton, M., & Midgley, C. (1997). Avoiding the demonstration of lack of ability: An underexplored aspect of goal theory. *Journal of Educational Psychology, 89,* 710-718.
Midgley, C. (1993). Motivation and middle level schools. In: M.L. Maehr & P.R. Pintrich (Eds), *Advances in Motivation and Achievement, Vol. 8: Motivation and Adolescent Development* (pp. 217-274). Greenwich, CT: JAI Press.
Midgley, C., Anderman, E., & Hicks, L. (1995). Differences between elementary and middle school teachers and students: A goal theory approach. *Journal of Early Adolescence, 15,* 90-113.
Midgley, C., & Edelin, K. (1998).Middle school reform and early adolescent well-being: The good news and the bad. *Educational psychologist, 33,* 195-206.
Midgley, C., Middleton, M. J., Gheen, M. H., & Kumar, R. (2002). Stage-environment fit revisited: A goal theory approach to examining school transitions. In: C.Midgley (Ed.), *Goals, Goal Structures, and Patterns of Adaptive Learning* (pp. 109-142). Mahwah, NJ: Erlbaum.
Midgley, C., Kaplan, A., Middleton, M., Maehr, M. L., Urdan, T., Anderman, L. H., et al. (1998). The development and validation of scales assessing students' achievement goal orientations. *Contemporary Educational Psychology, 23,* 113-131.
Midgley, C., & Urdan, T. C. (1992). The transition to middle level schools: Making it a good experience for all students. *Middle School Journal, 24,* 5-14.
Mook, D. G. (1986). *Motivation: The organization of action.* New York: Norton.
Mordkowitz, E.R., & Ginsburg, H.P. (1987). The academic socialization of successful Asian-American college students. *Quarterly Journal of Laboratory of Comparative Human Cognition, 9,* 85-91.
Mueller, C. M., & Dweck, C. S. (1998). Praise for intelligence can undermine children's motivation and performance. *Journal of Personality and Social Psychology, 75,* 33-52.
Multon, K.D., Brown, S.D., & Lent, R.W.(1991).Relations of self-efficacy beliefs to academic outcomes: A meta-analytic investigation. *Journal of Counseling Psychology, 38,* 30-38.
Murray, H. (1938). *Exploration in personality.* New York: Oxford University Press.
Muzzatti, B., & Agnoli, F.(2007).Gender and mathematics: Attitudes and stereotype threat susceptibility in Italian children. *Developmental Psychology, 43,* 747-759.
Neber, H., & Heller, K. A. (2002). Evaluation of a summer school program for highly gifted secondary school students: The German Pupils Academy. *European Journal of Psychological Assessment, 18,* 214-218.
Neisser, U. (1992). The development of consciousness and the acquisition of skill. In F. Kessel, P. M. Cole, & D. Johnson (Eds.), *Self and consciousness: Multiple perspectives* (pp.1-18). Hillsdale, NJ: Erlbaum.
Nicholls, J.(1984). Achievement motivation: Conceptions of ability, subjective experience, task choice, and performance. *Psychological Review, 91,* 328-346.
Nicholls, J. G. (1986, April). *Adolescents' conceptions of ability and intelligence.* Paper presented at the Annual Meeting of the American Educational Research Association, San Francisco.
Nicholls, J. G. (1989). *The competitive ethos and democratic education.* Cambridge, MA: Harvard University Press.
Nolen, S. (1988).Reasons for studying: Motivational orientations and study strategies. *Cognition and Instruction, 5,* 269-287.
Organization for Economic Co-operation and Development (OECD) (2004). Learning for tomorrow's world: First results from PISA 2003. Paris, France: OECD.
Oettingen, G. (1993, November). Cross-cultural perspectives on self-efficacy beliefs. Paper presented to the conference on Self-efficacy in Adaptation of Youth in Changing Societies. Marbach, Germany.
Osipow, S. H., & Fitzgerald, L. F.(1996). *Theories of career development* (4th edition). Allyn & Bacon.
Pajares, F.(1996).Self-efficacy beliefs in academic settings.*Review of Educational Research, 66,*543-578.

Pajares, F., & Kranzler, J. (1995). Self-efficacy beliefs and general mental ability in mathematical problem solving. *Contemporary Educational Psychology, 20*, 427-444.

Pajares, F., & Miller, M.D. (1994). Role of self-efficacy and self-concept beliefs in mathematical problem solving: A path analysis. *Journal of Educational Psychology, 86*, 193-203.

Panksepp, J. (1994).Emotional development yields lots of "stuff"...especially mind "stuff" that emerges from brain "stuff." In P. Ekman & R. J. Davidson (Eds.), *The nature of emotion: Fundamental questions* (pp. 367-368). New York: Oxford University Press.

Parke, R.D. (1994).Progress, paradigms, and unresolved problems:Recent advances in our understanding of children's emotions. *Merrill-Palmer Quarterly, 40,* 157-169.

Paris, S.G., Lipson, M.Y., & Wixson, K. (1983). Becoming a strategic reader. *Contemporary Educational Psychology, 8,* 293–316.

Paris, S. G., & Oka, E.(1986).Children's reading strategies, metacognition and motivation. *Developmental Review, 6,* 25-86.

Parke, R.D., MacDonald, K.B., Burks, V.M., Carson, J., Bhavnagri, N., Barth, J.M., & Beitel, A. (1989). Family and peer systems: In search of linkages. In K. Kreppner & R.M. Lerner (Eds.), *Family systems and life span development* (pp.65-91). Hillsdale, NJ: Erlbaum.

Parsons, J. E., Frieze, I. H., & Ruble, D. N. (1976). Introduction. *Journal of Social Issues, 32,* 1-5.

Parsons, J. E., Ruble, D. N., Hodges, K. L, & Small, A. W. (1976). Cognitive-developmental factors in emerging sex differences in achievement-related expectancies. *Journal of Social Issues, 32,* 47-61.

Patrick, H. (2004). Re-examining classroom mastery goal structure. *Advances in Motivation and Achievement, Vol. 13: Motivating students, improving schools.* (pp. 233-263). Elsevier Ltd.

Patrick, B. C., Skinner, E. A., & Connell, J. P. (1993). What motivates children's behavior and emotion? Joint effects of perceived control and autonomy in the academic domain.*Journal of Personality and Social Psychology, 65,* 781-791.

Patrick, H., Ryan, A. M., & Pintrich, P. R. (1999). The differential impact of extrinsic and mastery goal orientations on males' and females' self-regulated learning. *Learning and Individual Differences, 11,* 153-171.

Pekrun, R. (1988). *Emotion, Motivation und Persönlichkeit* (Emotion, motivation and personality). München, Germany: Psychologie Verlags Union.

Pekrun, R. (1992 c). Expectancy-value theory of anxiety: Overview and implications. In D.G. Forgays, T. Sosnowski, & K. Wrzesniewski (Eds.), *Anxiety: Recent developments in self-appraisal, psycho physiological and health research* (pp. 23-41). Washington, DC: Hemisphere.

Pekrun, R. (1998). Schüleremotionen und ihre Förderung: Ein blinder Fleck der Unterrichtsforschung (Students'emotions: A neglected topic of educational research). *Psychologie in Erziehung und Unterricht, 44,* 30-248.

Pekrun, R.(2000). A social-cognitive, control-value theory of achievement emotions. In J. Heckhausen (Ed.), *Motivational psychology of human development.* Oxford, UK: Elsevier.

Pekrun, R. (2006). The control-value theory of achievement emotions: Assumptions, corollaries, and implications for educational research and practice. *Educational Psychology Review, 18,* 315-341.

Pekrun, R., Elliot, A.J., & Maier, M.A. (2006). Achievement goals and discrete achievement emotions: A theoretical model and prospective test. *Journal of Educational Psychology, 98,* 583-597.

Pekrun R., & Hofmann, H. (1999). Lern- und Leistungsemotionen: Erste Befunde eines Forschungsprogramms (Emotions in learning and achievement: First results of a program of research). In M. Jerusalem & R. Pekrun (Eds.), *Emotion, Motivation und Leistung* (pp. 247-267). Göttingen, Germany: hogrefe.

Pekrun, R., Goetze, T., Titz, W., & Perry, R.P. (2002a). Academic emotions in students' self-regulated learning and achievement: A program of qualitative and quantitative research. *Educational Psychologist, 37,* 91-106.

Pekrun, R., Götz, T., Jullien, S., Zirngibl, A., v. Hofe, R., & Blum, W. (2002b). Skalenhandbuch PALMA:1.Messzeitpunkt(5.Klassenstufe). Universität München:Institut Pädagogische Psychologie.

Pekrun, R., Goetz, T., & Perry, R. P. (2005a). *Academic Emotions Questionnaire (AEQ). User's manual.* Department of Psychology, University of Munich.

Pekrun, R., Goetz, T., & Frenzel, C.(2005b).*Academic Emotions Questionnaire – Mathematics (AEQ-M). User's manual.* Department of Psychology, University of Munich.

Perry, R. P., Schonwetter, D., Magnusson, J-L., & Struthers, W. (1994). Use of explanatory schemas and the quality of college instruction: Some evidence for buffer and compensation effects. *Research in Higher Education, 35*, 349-371.
Piaget, J. (1952). *The origins of intelligence in children.* New York: Routledge & Kagan Paul.
Piaget, J. (1962). *Play, dreams and imitation in childhood.* New York: Norton.
Piaget, J. (1967). *Six psychological studies.* New York: Random House.
Piaget, J., & Inhelder, B. (1969). *The psychology of the child.* New York: Basic.
Pintrich, P.R., Smith, D.A., Garcia, T., & McKeachie, W.J.(1991).*A manual for the use of the Motivated Strategies for Learning Questionnaire (MSLQ).* Ann Arbor: University of Michigan, National Center for Research to Improve Postsecondary Teaching and Learning.
Pintrich, P.R., Smith, D.A., Garcia, T., & McKeachie, W. J. (1993). Reliability and predictive validity of the Motivated Strategies for Learning Questionnaire (MSLQ). *Educational and Psychological Measurement, 53*, 801-813.
Pintrich, P. R. (1988). A process-oriented view of student motivation and cognition. In J. S. Stark & L. Mets (Eds.), *Improving teaching and learning through research. New directions for institutional research, 57* (pp.55-70). San Francisco: Jossey-Bass.
Pintrich, P. R. (1989). The dynamic interplay of student motivation and cognition in the college classroom. In C. Ames & M. Maehr (Eds.), *Advances in motivation and achievement: Vol. 6. Motivation enhancing environments* (pp. 117–160). Greenwich, CT: JAI Press.
Pintrich, P. (2000). The role of goal orientation in self-regulated learning. In M. Boekaerts, P. Pintrich & M. Zeidner (Eds.), *Handbook of self-regulation: Theory, research and applications* (pp. 451-502). San Diego, CA: Academic Press.
Pintrich, P. R., Cross, D. R., Kozma, R.B., & McKeachie, W.J. (1986). Instructional psychology. *Annual Review of Psychology, 37*, 611–651.
Pintrich, P.R., McKeachie,W.J., & Lin,Y.G. (1987). Teaching a course in learning to learn. *Teaching of Psychology, 14*, 81-86.
Pintrich, P.R., &De Groot, E.V.(1990).Motivational and Self-Regulated Learning Components of Classroom Academic Performance. *Journal of Educational Psychology, 82*, 33 – 40.
Plutchik, R. (1983). Emotions in early development: A psychoevolutionary approach. In R. Plutchik & H. Kellerman (Eds.), *Emotions: Theory, research, and experience* (pp.221-257). New York: Academic.
Purkey,W.(1978).Inviting school success: *A self-concept approach to teaching and learning.* Englewood Cliffs, NJ: Prentice-Hall.
Purkey, W. (1990, February). Self-esteem: On the wings of turtles. Address given at the California Conference on Self-Esteem, San Jose, CA.
Quay, H. C. (1986). Classification. In H. C. Quay / J. S. Werry (Eds.), *Psychopathological disorders of childhood* (3rd ed.). New York: Wiley.
Quinn, D. M., & Spencer, S. J. (2001). The interference of stereotype threat with women's generation of mathematical problem-solving strategies. *Journal of Social Issues, 57*, 55-71.
Rao, N., & Sachs, J. (1999). Confirmatory factor analysis of the Chinese version of the Motivated Strategies for Learning Questionnaire. *Educational and Psychological Measurement, 59*, 1016-1029.
Reasoner, R. W. (1982). *Building self- esteem: A comprehensive program.* Palo Alto, CA: Consulting Psychologists Press.
Reasoner,R.W.(1992).*Building self-esteem in the elementary schools.*ConsultingPsychologistsPress, Inc.
Reasoner, R. W. (1994). *Building self-esteem in the elementary schools:Administrator's guide.* (2nd Ed.). Consulting Psychologists Press, Inc.
Reid, D. K. (1988). *Teaching the Learning Disabled: A Cognitive Developmental Approach.* Boston: Allyn & Bacon.
Renninger, K. A. (2000). Individual interest and its implications for understanding intrinsic motivation. In C. Sansone & J. M. Harackiewicz (Eds.), *Intrinsic and extrinsic motivation: The search for optimal motivation and performance* (pp. 373-404). San Diego, CA: Academic Press.
Riesemberg, R., & Zimmerman, B.J. (1992). Self-regulated learning in gifted students. *Roeper Review, 15*, 98-101.
Robinson, A. (1990). Does that describe me? Adolescents' acceptance of the gifted label. *Journal of the Education of the Gifted, 13*(3), 245-255.

Roeser, R. W., Midgley, C., & Urdan, T.C. (1996). Perceptions of the school psychological environment and early adolescents' psychological and behavioral functioning in school: The mediating role of goals and belonging. *Journal of Educational Psychology, 88,* 408-422.
Rosaldo, M. Z. (1984). Toward an anthropology of self and feeling. In R.A. Schweder, & R. A. LeVine (Eds.), *Cultural theory: Issues on mind, self, and emotion* (pp. 137-157). Cambridge, England: Cambridge University Press.
Roseman, I. J. (1984). Cognitive determinants of emotion. *Review of Personality and Social Psychology, 5,* 11-36.
Rosenbaum, R. M.(1972).*A dimensional analysis of the perceived causes of success and failure.* Unpublished doctoral dissertation, University of California, Los Angeles.
Rotter, J. B. (1966). Generalized expectancies for internal versus external control of reinforcement. *Psychological Monograph, 80,* 1-28.
Rosen, B.C., & D'Andrade, R. (1975).The psycho-social origins of achievement motivation. In Bronfenbrenner. U., & Mahoney, M. A. (Eds.). *Influences on human development.* The Dryden Press: Hinsdale, Illinois.
Rothbart, M. K., Posner, M. I., & Hershey, K. L. (1995). Temperament, attention and developmental psychopathology. In D. Cicchetti & D. Cohen (Eds.), *Manual of developmental psychopathology* (pp.315-340). New York: Wiley.
Ryan, N. M. (1989). Stress-coping strategies identified from school age children's perspective. *Research in Nursing and Health, 20,* 111-122.
Ryan, R. M., & Deci, E. L. (2000).Intrinsic and extrinsic motivations: Classic definitions and new directions. *Contemporary Educational Psychology, 25,* 54-67.
Ryan, A. M., Hicks, L., & Midgley, C. (1997). Social goals, academic goals, and avoiding seeking help in the classroom. *Journal of Early Adolescence, 17,* 152-171.
Ryan, A.M., & Patrick, H. (2001). The classroom social environment and changes in adolescents' motivation and engagement during middle school. *American Educational Research Journal, 38,* 437-460.
Ryan, A.M., & Pintrich, P.R. (1997). "Should I ask for help?" The role of motivation and attitudes in adolescents' help seeking in math class. *Journal of Educational Psychology, 89,* 329-341.
Saarni, C. (1990). Emotional competence: How emotions and relationships become integrated. In R. A. Thompson (Ed.), *Socioemotional development* (pp.115-182). Lincoln: University of Nebraska Press.
Saarni, C. (1997). Emotional competence and self-regulation in childhood. In *Emotional development and emotional intelligence: Educational implications.* Salovey, P., & Sluyter, D. J. (Eds.). Basic Books: A Division of Harper Collins Publishers.
Salili, F. (1994). Age, sex, and cultural differences in the meaning and dimensions of achievement. *Personality and Social Psychology Bulletin, 2,* 635-648.
Salili, F. (1995). Explaining Chinese students' motivation and achievement: A socio-cultural analysis. *Advances in Motivation and Achievement, 9,* 73-118. JAI Press Inc.
Salili, F., & Ching, S. L. (1992)*Antecedents of motivation and achievement among the Chinese.* Unpublished research report, University of Hong Kong.
Salili, F., & Ho, W. (1992).*Motivation and achievement: A Chinese study.* Unpublished research report, University of Hong Kong.
Salili, F., Hwang, C. E., & Choi, N.F. (1989). Teachers' evaluative behavior: The relationship between teachers' comments and perceived ability in Hong Kong. *Journal of Cross-Cultural Psychology, 20*(2), 115-132.
Sansone, C., & Smith, J.L. (2000). Interest and self-regulation:The relation between having to and wanting to. In C, Sansone & J. M. Harackiewicz. (Eds.), *Intrinsic and extrinsic motivation: The search for optimal motivation and performance* (pp. 341-372). San Diego, CA: Academic Press.
Santrock, J. (1990). *Adolescence.* Dubuque, IA: Brown.
Schachter, S. (1954). Interpretative an methodological problem of replicated research. *Journal of Social Issues, 10,* 52-60.
Scherer, K. R. (1984). On the nature and function of emotion: A component process approach. In K. R. Scherer & P. Ekman (Eds.), *Approaches to emotion* (pp. 293-317). Hillsdale, NJ: Erlbaum.
Schiefele, U. (1991). Interest, learning, and motivation. *Educational Psychologist, 26,* 299-323.
Schore, A.N. (1994).*Affect regulation and the origin of the self: The neurobiology of emotional development.* Hillsdale, NJ: Erlbaum.

Schrest, L.M., Fay, T., & Zaide, S.(1972). Problems of translation in cross-cultural research. *Journal of Cross-Cultural Psychology, 1,* 41-56.
Schunk, D. H. (1983). Ability versus effort attributional feedback: Differential effects on self-efficacy and achievement. *Journal of Educational Psychology, 75,* 848-856.
Schunk, D. H. (1984). Self-efficacy perspective on achievement behavior. *Educational Psychologist, 19,* 48-58.
Schunk, D. H. (1985). Self-efficacy and school learning. *Psychology in the Schools, 22,* 208-223.
Schunk, D. H. (1987, September). *Self-efficacy and cognitive achievement.* Paper presented at the annual meeting of the American Psychological Association, New York.
Schunk, D. H. (1991). Self-efficacy and academic motivation. *Educational Psychologist, 26,* 207-231.
Schunk, D. H. (1994). Self-regulation of self-efficacy and attributions in academic settings. In D. H. Schunk & B. J. Zimmerman (Eds.), *Self-regulation of learning and performance: Issues and educational implications* (pp. 75-99). Hillsdale, NJ: Erlbaum.
Schunk, D. H. (1986). Strategy training and attributional feedback with learning disabled students. *Journal of Educational Psychology, 78,* 201-209.
Schunk, D. H., & Zimmerman, B. J. (1994). *Self-regulation of learning and performance: Issues and educational applications.* Hillsdale, NJ: Erlbaum.
Schutz, P. A. (1994). Goals as the transactive point between motivation and cognition. In P. R. Pintrich, D. Brown, & C. E. Weinstein (Eds.), *Perspectives on student motivation, cognition, and learning: Essays in honor of Wilbert J. McKeachie* (pp. 113-133). Hillsdale, NJ: Erlbaum.
Schutz, P. A., Crowder, K. C., & White, V. E. (2001). The development of a goal to become a teacher. *Journal of Educational Psychology, 93,* 299-308.
Schutz, P. A., & Davis, H.A. (2000). Emotions and self-regulation during test taking. Educational Psychologist, 35, 243-255.
Schutz, P. A., & DeCuir, J. T. (2002). Inquiry on emotions in education. *Educational Psychologist, 37,* 125-134.
Schutz, P. A., Hong, J. Y., Cross, D. I., & Osbon, J. N. (2006). Reflections on investigating emotion in educational activity settings. *Educational Psychology Review, 18,* 343-360.
Schweder, R. A., & Bourne, E. J. (1984). Does the concept of the person vary cross-culturally? In R.A. LeVine (Eds.), *Culture theory: Essays on mind, self and emotion* (pp. 158-199).Cambridge, England: Cambridge University Press.
Scott, G. J. (1983). Career Search, Selection, and Entry. In Walsh W. B., & Osipow, S. H. (Eds.), *Handbook of vocational Psychology, Vol. 2,* 77-97.. Lawrence Erlbaum Association, Inc.
Shih, S.S. (2005).Taiwan sixth graders' achievement goals and their motivation, strategy use, and grades: An examination of the multiple goal perspective. *Elementary School Journal, 106,* 39-58.
Shih, M, Pittinsky, L., & Ambady, N. (1999). Stereotype susceptibility: Identity salience and shifts in quantitative performance. *Psychological Science, 10,* 80-83.
Simmons, R. G., & Blyth, D. A. (1987). *Moving into adolescence.* Hawthorne, NY: de Gruyter.
Simpkins, S.D., Davis-Kean, P.E., & Eccles, J.S. (2006). Math and science motivation: A longitudinal examination of the links between choices and beliefs. *Developmental Psychology, 42,* 70-83.
Singelis, T.M., & Brown, W. J. (1995). Culture, self, and collectivist communication: Linking culture to individual behavior. *Human Communication Research, 21,* 354-389.
Singh, R. (1981). Prediction of performance from motivation and ability: An appraisal of a cultural difference hypothesis. In J. Pandey (Ed.), *Perspectives on experimental social psychology in Indian.* New Delhi: Concept.
Skaalvik, E. M. (1997). Self-enhancing and self-defeating ego orientation: Relations with task and avoidance orientation, achievement, self-perceptions, and anxiety. *Journal of Educational Psychology, 89,* 71-81.
Skolnick, A.(1986). *The Psychology of Human Development.* San Diego: Harcourt, Brace, Jovanovich.
Smith, C.P. (Ed.). (1969). *Achievement-related motives in children.* Russell Sage Foundation, New York.
Smith, C.A., & Ellsworth, P.C. (1985).Patterns of cognitive appraisal in emotion. *Journal of Personality and Social Psychology, 48,* 813-838.
Snow, R. (1989). Aptitude-treatment interaction as a framework for research on individual differences in learning. In P. Ackerman, R.Sternberg, & R.Glaser (Eds.), Learning and individual differences (pp. 13-59). New York: Freeman.

Snow, R.E., Corno, L., & Jackson, D.N.(1996).Individual differences in affective and conative functions. In D. C. Berliner & R. C. Calfee (Eds.), *Handbook of educational psychology* (pp.243-310). New York: Simon & Schuster.
Sobel, M. E. (1982). Asymptotic confidence intervals for indirect effects in structural equation models. In S. Leinhart (Ed.), Sociological methodology. 1982, pp. 290-312. San Francisco: Jossey-Bass.
Solomon, R. C. (1984). Getting angry: The Jamesian theory of emotion in anthropology. In R. A. Shweder & R. A. LeVine (Eds.), *Culture theory: Essays on mind, self, and emotion* (pp. 238-254). Cambridge, England: Cambridge University Press.
Spelke, E. S. (2005). Sex differences in intrinsic aptitude for mathematics and science? A critical review. *American Psychologist, 60,* 950-958.
Spelke, E. S., & Grace, A. D. (2006). Abilities, motives, and personal styles. *American Psychologist, 61,* 725-726.
Sroufe, L. A. (1989). Pathways to adaptation and maladaptation: Psychopathology as developmental deviation. In D. Cicchetti (Ed.), *Rochester Symposia on Developmental Psychopathology* (Vol.1, pp. 13-14). Hillsdale, NJ:Erlbaum.
Sroufe, L.A. (1996).*Emotional development: The organization of emotional life in the early years.* Cambridge University Press.
Stern, D. (1985). *The interpersonal world of the infant: A view from psychoanalysis and developmental psychology.* New York: Basic.
Stevenson, H.W., & Lee, S. (1990). Context of Achievement, Monographs of the Society for Research in *Child Development, Serial no. 221, Vol. 55,* Nos. 1-2.
Stigler, J. W., Smith, S., & Mao, L. W. (1985). The self-perception of competence by Chinese children. *Child Development, 56,* 1259-1270.
Stipek, D. J. (1983). A developmental analysis of pride and shame. *Human Development, 26,* 42-54.
Stipek, D. J., & Gralinski, J. H. (1996). Children's beliefs about intelligence and social performance. *Journal of Educational Psychology, 88,* 397-407.
Stipek, D., Weiner, B., & Li, K. (1989). Testing some attribution-emotion relations in the People's Republic of China. *Journal of Personality and Social Psychology, 56,* 109-116.
Stöber, J., & Pekrun, R. (2004). Advances in test anxiety research. *Anxiety, Stress, and Coping, 17,* 205-211.
Spence, J. T., & Helmreich, R. L.(1983). Achievement-related motives and behaviors. In *Achievement and achievement motives: psychological and sociological approaches. Spence, J. T. (Ed.).W. H. Freeman and Company San Francisco.*
Stapley, J. C., & Haviland, J. M.(1989). Beyond depression: Gender differences in normal adolescents' emotional experiences. *Sex Roles, 20,* 295-308.
Steele, C. M., Spencer, S. J., & Aronson, J. (2002). Contending with group image: The psychology of stereotype and social identity threat. *Advances in Experimental Social Psychology, 34,* 379-440.
Sternberg, L. (1990). Autonomy, conflict, and harmony in the family relationship. In S. Feldman & G. Elliott (Eds.), *At the threshold:The developing adolescent* (pp. 255-276). Cambridge, MA: Harvard University Press.
Sullivan, H. S. (1947). *Conceptions of modern psychiatry.* Washington: Wm. Allanson White Psychiat.
Super, D. E., & Bohn, M. J., Jr. (1971). *Occupational Psychology.* Tavistock Publications.
Sutton, R. E. (2004). Emotional regulation goals and strategies of teachers. *Social Psychology of Education, 7,* 379-398.
Terman, L. M., & Merrill, M. A. (1937). *Measuring intelligence.* Boston: Houghton Mifflin.
Thatcher, R.W.(1994).Psychopathology of early frontal lobe damage: Dependence on cycles of development. *Development and psychopathology, 6,* 565-596.
Thompson, R. A. (1990). Emotion and self-regulation. *Nebraska Symposium on Motivation* (pp. 367-467).
Tiedemann, J. (2000). Parents' gender stereotypes and teachers' beliefs as predictors of children's concept of their mathematical ability in elementary school. *Journal of Educational Psychology, 92,* 144-151.
Tobias, S. (1978). *Overcoming math anxiety.* New York: Norton.
Tobias, S. (1985). Test anxiety: Interference, defective skills, and cognitive capacity. Educational Psychologist, 20, 135-142.

Trafimow, D., Triandis, H. C., & Goto, S. G. (1991). Some tests of the distinction between the private self and the collective self. *Journal of Personality and Social Psychology, 60,* 649-655.

Tresemer, D. (1977). *Fear of success.* New York: Plenum.

Triandis, H. C. (1972). *The analysis of subjective culture.* New York: Wiley.

Triandis, H. C. (1997). Cross-cultural perspective on personality. In R.Hogan, J.Johnson, & S.R. Briggs (Eds.), *Handbook of personality psychology* (pp. 439-464). San Diego, CA: Academic Press.

Triandis, H. C. (1989).The self and social behavior in differing cultural contexts.*Psychological Review, 96,* 506-520.

Triandis, H. C. (1995). Motivation and achievement in collectivist and individualist cultures. *Advances in Motivation and Achievement, 9,* 1-30. JAI Press Inc.

Tucker, D. M. (1992). Developing emotions and cortical networks. In M. R. Gunnar & C. A. Nelson (Eds.), *Minnesota Symposia on Child Psychology: Vol. 24, Developmental behavioral neuroscience* (pp. 75-128). Hillsdale NJ: Erlbaum.

Updegraff, K. A., Eccles, J. S., Barber, B. L., & O'Brien, K. M. (1996). Course enrollment as self-regulatory behavior: Who take optional high school math courses? *Learning and Individual Differences, 8,* 239-259.

Urdan, T. (1997). Achievement goal theory: Past results, future directions. In M. Maehr & P. Pintrich (Eds.), Advances in motivation and achievement pp.99-141). Greenwich, CT: JAI.

Urdan, T. C. & Roeser, R. W. (1993). *The relations among adolescents' social cognitions, affect, and academic self-schemas.* Paper presented at the Annual Meeting of the American Educational Research Association, Atlanta.

Vallerand, Gagné, F., Senécal, C., & Pelletier, L.G. (1994). A comparison of the school intrinsic motivation and perceived competence of gifted and regular students.*Gifted Child Quarterly, 38,*172-175.

Vandewalle, D. (1997). Development and validation of a work domain goal orientation instrument. *Educational and Psychological Measurement, 57,* 995-1015.

Van Gennep, A. (1960). *The rites of passage.* Chicago: Chicago University Press.

Veroff, J., McClelland, L., & Ruhland, D. (1975). Varieties of achievement motivation. In M.T.S. Mednick, Tangri, S.S., and Hoffman, L.W. (Eds.). *Women and achievement: Social and motivational analyses.* Washington, DC: Hemisphere.

Vygotsky, L. (1978). *Mind and society.* Cambridge, MA: Harvard University Press.

Waldrop, W. M.(1992). *Complexity.* New York: Simon & Schuster.

Wan, Y. T.(2004).Emotions in learning mathematics in China and Germany: A cross-cultural study. Unpublished master thesis. University of Munich, Munich, Germany.

Waston, R.I., & Lindgren, H.C. (1979). *Psychology of the child and the adolescent.* Macmillan Publishing Co., Inc. New York.

Weiner, B. (1972). *Theories of motivation: From mechanism to cognition.* Chicago: Markhan.

Weiner, B. (1974). *Achievement motivation and attribution theory.*Morriston,NJ:General Learning Press.

Weiner, B. (1979). A theory of motivation for some classroom experiences. *Journal of Educational Psychology, 71,* 3-25.

Weiner, B. (1980a). A cognitive (attribution)-emotion-action model of motivated behavior. An analysis of judgments of help-giving. *Journal of Personality and Social Psychology, 39,* 186-200.

Weiner, B. (1980b). May I borrow your class notes? An attributional analysis of judgments of help-giving in an achievement related context. *Journal of Educational Psychology, 72,* 676-681.

Weiner, B. (1983). Some methodological pitfalls in attributional research. *Journal of Educational Psychology, 75,* 530-543.

Weiner, B. (1985). An attribution theory of achievement motivation and emotion. *Psychological Review, 92,* 548-573.

Weiner, B. (1986). *An attributional theory of motivation and emotion.* New York: Springer-Verlag.

Weiner, B. (1994). Ability versus effort revisited: The moral determinants of achievement evaluation and achievement as a moral system. *Educational Psychologist, 29,* 163-172.

Weiner, B., Frieze, I. H., Kukla, A., Reed, L., Rest, S., & Rosenbaum, R. M. (1971). *Perceiving the causes of success and failure.* Morristown, NJ: General Learning Press.

Weiner, B., Graham, S., & Chandler, C. (1982). Causal antecedents of pity, anger and guilt.*Personality and Social Psychology Bulletin, 8,* 226-232.

Weiner, B., Russell, D., & Lerman, D. (1978). Affective consequences of causal ascriptions. In J. H. Harvey, W. J. Ickes, & R. F. Kidd, (Eds), *New directions in attribution research* (Vol. 2. pp. 59-88). Hillsdale, NJ: Erlbaum.
Weiner, B., Russell, D., & Lerman, D. (1979). The cognition-emotion process in achievement-related contexts. *Journal of Personality and Social Psychology, 37,* 1211-1220.
Weinert, F. (1987). Metacognition and motivation as determinants of effective learning and understanding. In F. Weinert & R. Kluwe (Eds.), *Metacognition, motivation, and understanding* (pp. 1-15). Hillsdale, NJ: Erlbaum.
Weinert, F. E., & Helmke, A. (1995). Learning from wise mother nature or big brother instructor: The wrong choice as seen from an educational perspective. *Educational Psychologist, 30,* 135-142.
Weinstein, R. S., & Butterworth, B. (1993, April). *Enhancing motivational opportunity in elementary schooling: A case study of the principal's role.* Paper presented at the Annual Meeting of the American Educational Research Association, Atlanta.
Weinstein, C.E., & Mayer, R.E. (1986). The teaching of learning strategies. In M. Wittrock (Ed.), Handbook of research on teaching (pp. 315-327). New York: Macmillan.
Weisfeld, G.E. (1997). Puberty rites as clues to the nature of human adolescence. *Cross-Cultural Research, 31(1),* 27-54.
Weisfeld, G. E. (1999). *Evolutionary principles of human adolescence.* New York: Basic Books.
Werner, H., & Kaplan, B. (1963). *Symbol formation: An organismic-developmental approach to language and the expression of thought.* New York: Wiley.
Werner, O., & Campbell, D.T. (1970). Translating, working through interpreters, and the problem of decentering. P. 398-399. In Naroll, R., & Cohen, R. (Ed), *A handbook of method in cultural anthropology.*
Westermeyer, J. (1987). Cultural factors in clinical assessment. *Journal of consulting and Clinical Psychology, 55,* 471-478.
White, R.W. (1963). Ego and reality in psychoanalytic theory: A proposal regarding independent ego energies; *Psychological Issues Monograph* (No. 11). New York: International Universities Press.
Whiteley, J. M. (1978). Career counseling: An overview, In J.M. Whiteley & A. Resnikoff (Eds.), *Career counseling.* Monterey, Calif.: Brooks/Cole.
Wicker, F. W., Payne, G. C., & Morgan, R. D. (1983). Participant descriptions of guilt and shame. *Motivation and Emotion, 7,* 25-39.
Wigfield, A., & Eccles, J. (1989). Test anxiety in elementary and secondary school students. *Educational Psychologist, 24,* 159-183.
Wigfield, A., Eccles, J.S., Mac Iver, D., Reuman, D.A., & Midgley, C. (1991). Transitions during early adolescence: Changes in children's domain-specific self-perceptions and general self-esteem across the transition to junior high school. *Developmental Psychology, 27,* 552-565.
Wigfield, A., & Karpathian, M. (1991). Who am I and what can I do? Children's self-concepts and motivation in achievement situations. *Educational Psychologist, 26,* 233-261.
Wilson, T.D., & Linville, P.W. (1982). Improving the academic performance of college freshmen: Attribution theory revisited. *Journal of Personality and Social Psychology, 42,* 367-376.
Wilson, T. D., & Linville, P.W. (1985).Improving the performance of college freshmen with attributional techniques. *Journal of Personality and Social Psychology, 49,* 287-293.
Wilson, R.W., & Pusey, A.W. (1982). Achievement motivation and small business relationship patterns in Chinese society. In S. L. Greenblatt, R.W. Wilson, & A.A. Wilson (Eds.), *Social Interaction in Chinese Society.* (pp. 195-208). New York: Praeger.
Winston, M., Vahala, M., Nichols, E., Gillis, M., Winthrow, M., & Rome, K.(1994). A measure of college classroom climate. *Journal of College Student Development, 35,* 11-18.
Winter, S. (1990).Teacher approval and disapproval in Hong Kong secondary school classrooms. *British Journal of Educational Psychology, 60,* 88-92.
Wolters, C. A., Yu, S. L., & Pintrich, P. R. (1996). The relation between goal orientation and students' motivational beliefs and self-regulated learning. *Learning and Individual Difference, 8,* 211-238.
Yang, K. S. (1986). Chinese personality and its change. In Bond, M. H. (Ed.), *The Psychology of the Chinese People* (pp. 106-160). Hong Kong: Oxford University Press.
Ye, H. Z. (2004).Emotions in learning mathematics in China and Germany – A cross-cultural study. Unpublished master thesis. University of Munich, Munich, Germany.

Yerkes, R. M., Bridges, J. W., & Hardwick, R. S. (1915). *A point scale for measuring mental ability.* Baltimore: Warwick & York.

Yu, E.S.H.(1974).Achievement motive, familism, and hsiao: A replication of McClelland-Winterbottom studies. *Dissertation Abstracts International, 35*, 593A (University Microfilms, No. 74-14, 9442).

Yu, A. B., & Yang, K. S. (1987). Social-and individual-oriented achievement motivation: A conceptual and empirical analysis. *Bulletin of the Institute of Ethnology, Academica Sinica* (Taipei, Taiwan), 64, 51-98. (In Chinese).

Zajonc, R. (1998). Emotion. In D. Gilbert, S. Fiske, & G. Lindzey (Eds.), *The handbook social psychology, 4th edition* (pp.591-632). New York: McGraw-Hill.

Zeidner, M. (1998). *Test anxiety: The state of the art.* Plenum, New York.

Zeidner, M., & Endler, N. (Eds.).(1996). *Handbook of coping:Theory, research, applications.* New York: Wiley.

Zimmerman, B., & Pons, M. (1986). Development of a structured interview for assessing student use of self-regulated learning strategies. *American Educational Research Journal, 23*, 614-628.

Zimmerman, B., & Pons, M. (1988). Construct validation of a strategy model of student self-regulated learning. *Journal of Educational Psychology, 80*, 284–290.

Zimmerman, B.J., Bandura, A., & Martinez-Pons, M. (1992). Self-motivation for academic attainment: The role of self-efficacy beliefs and personal goal setting. *American Educational Research Journal, 29, 663-676.*

Zoeller, C., Mahoney, G., & Weiner, B. (1983). Effects of attribution training on the assembly task performance of mentally retarded adults. *American Journal of Mental Deficiency, 88*, 109-112.

www.ingramcontent.com/pod-product-compliance
Ingram Content Group UK Ltd.
Pitfield, Milton Keynes, MK11 3LW, UK
UKHW020857160426
5217IPUK00035B/1353